Lectures in Mathematics
ETH Zürich
Department of Mathematics
Research Institute of Mathematics

J. Donald Monk

Cardinal Functions on Boolean Algebras

1990

Birkhäuser Verlag
Basel · Boston · Berlin

Author's address:

J. Donald Monk
Mathematics Department
University of Colorado
Boulder, CO 80309-0426
USA

Library of Congress Cataloging-in-Publication Data

Monk, J. Donald (James Donald). 1930 –
Cardinal functions on Boolean algebras / J. Donald Monk.
p. cm.
«Lectures in mathematics ETH Zürich»
Includes bibliographical references (p.) and indexes.
ISBN 3-7643-2495-3 (alk. paper) — ISBN 0-8176-2495-3 (alk. paper)
1. Algebra, Boolean. 2. Functions. 3. Numbers, Cardinal.
I. Title.
QA10.3.M66 1990
511.3'24–dc20

Deutsche Bibliothek Cataloging-in-Publication Data

Monk, James Donald:
Cardinal functions on Boolean algebras / J. Donald Monk. – Basel; Boston; Berlin:
Birkhäuser 1990
(Lectures in mathematics)
ISBN 3-7643-2495-3 (Basel ...) brosch.
ISBN 0-8176-2495-3 (Boston) brosch.

Printed in Germany on acid-free paper
ISBN 3-7643-2495-3
ISBN 0-8176-2495-3

To Hans and Eli

FOREWORD

This is a somewhat revised version of a *Nachdiplomvorlesung* given at the Forschungsinstitut für Mathematik, ETH Zürich, during the summer semester of 1987. Since giving the lectures I have continued to revise and expand the original notes. The handwritten notes produced by Andrea G. Clivio for my lectures are omitted here. They covered some background material, and added details in some of the proofs. The background material is available now in the Handbook of Boolean Algebras (North-Holland, 1989); only volume 1 of that work is needed, and in fact only parts of it. A reader unversed in the theory of Boolean algebras is advised to read that volume before starting with these notes; the following (sub)sections can be skipped: 3.3, 4.4, 5.3, 5.5, 5.6, 6.2, 8.2, 8.3, 10.2, 10.3, 11.2, 12.2, 12.3, 12.4, all of sections 13 and 14, 16.3, 17.3, and all of Chapter 7. Some of the proofs given in the original lectures have been expanded.

These notes are still in a preliminary form. Many specific problems and non-specific queries are stated. The reader should bear in mind that these problems and queries may already be answered somewhere in the literature. Also, while I have attempted to be accurate about historical matters and to give sufficient references for the reader to be able to track down relevant results, the historical remarks are tentative.

For helpful comments (in many cases, solutions of problems stated in earlier versions of these notes, or indications of relevant literature) I am indebted to E. K. van Douwen, L. Heindorf, S. Koppelberg, P. Koszmider, P. Nyikos, S. Shelah, and S. Todorčević. This is not to say that any of these kind people are responsible for shortcomings in this book. In fact, when one of them pointed out something to me (solution of a question, or something), that usually led to some additional questions, and I again circulated a version of these notes; this iteration was repeated many times. As a result, the final version contains many statements and problems not seen by them. Therefore it is likely that some of the questions here are trivial, naive, known in the literature, or otherwise ill-considered.

J. Donald Monk
Boulder, Colorado
March 7, 1990

CONTENTS

0. Introduction 1
1. Cellularity 7
2. Depth 32
3. Topological density 41
4. π-weight 47
5. Length 53
6. Irredundance 57
7. Cardinality 66
8. Independence 69
9. π-character 76
10. Tightness 82
11. Spread 89
12. Character 94
13. Hereditary Lindelöf degree 99
14. Hereditary density 103
15. Incomparability 105
16. Hereditary cofinality 111
17. Number of ultrafilters 117
18. Number of automorphisms 118
19. Number of endomorphisms 120
20. Number of ideals 122
21. Number of subalgebras 123

Diagrams 127
Examples 134
References 140
Index of problems 145
Index of names 149
Index of symbols 150
Index of words 151

0. INTRODUCTION

These notes are concerned with the theory of the most common functions k which assign to each Boolean algebra A a cardinal number kA. Examples of such functions are the cardinality of the algebra A, and $\sup\{|X| : X$ is a family of pairwise disjoint elements of $A\}$. We have selected 21 such functions as the most important ones, and several others are mentioned as we go along. Let us mention right away an ambiguity in these notes: *Many statements are valid only if the Boolean algebras considered are infinite.* For each function one can consider two very general questions: (1) How does the function behave with respect to algebraic operations, e.g., what is the value of k on a subalgebra of A in terms of its value on A? (2) What can one say about other cardinal functions naturally derived from a given one, e.g., what is $\sup\{kB : B$ is a homomorphic image of $A\}$? Another very general kind of question concerns the relationship between the various cardinal functions. We shall shortly be more specific about what these three general questions amount to. The purpose of these notes is to survey this area of the theory of BA's, giving proofs for a large number of results, some of which are new, mentioning most of the known results, and formulating open problems. The open problems are of two sorts: very definite problems, which will be enumerated, and indefinite problems—frequently concerning functions naturally arising that have evidently not been investigated—which will be mentioned only informally.

The framework that we shall set forth and then follow in investigating cardinal functions seems to us to be important for several reasons. First of all, the functions themselves seem intrinsically interesting. Many of the questions which naturally arise can be easily answered on the basis of our current knowledge of the structure of Boolean algebras, but some of these answers require rather deep arguments of set theory, or of topology. This provides another interest in their study: as a natural source of applications of set-theoretical or topological methods. Many of the questions are unresolved. Some of these are rather obscure and uninteresting, but some of them have a general interest. Altogether, the study of cardinal functions seems to bring a unity and depth to many isolated investigations in the theory of BA's.

There are several surveys of cardinal functions on Boolean algebras, or, more generally, on topological spaces: See Arhangelskiĭ [78], Comfort [71], van Douwen [89], Hodel [84], Juhász [71], Juhász [80], Juhász [84], and Monk [84]. We shall not assume any acquaintance with these articles. On the other hand, we shall frequently refer to results proved in Part I of the Handbook of Boolean Algebras.

Definition of the cardinal functions considered.

1. **Cellularity**. A subset X of a BA A is called *disjoint* if its members

are pairwise disjoint. The *cellularity* of A, denoted by $\mathrm{c}A$, is

$$\sup\{|X| : X \text{ is a disjoint subset of } A\}.$$

2. **Depth**. Depth$A$ is

$$\sup\{|X| : X \text{ is a subset of } A \text{ well-ordered by the Boolean ordering}\}.$$

3. **Topological density**. The *density* of a topological space X, denoted by $\mathrm{d}X$, is the smallest cardinal κ such that X has a dense subspace of cardinality κ. The *topological density* of a BA A, also denoted by $\mathrm{d}A$, is the density of its Stone space Ult$A$.

4. **π-weight**. A subset X of a BA A is *dense* in A if for all $a \in A^+$ there is an $x \in X^+$ such that $x \leq a$. The *π-weight* of a BA A, denoted by πA, is the smallest cardinal κ such that A has a dense subset of cardinality κ. This could also be called the *algebraic density* of A.

5. **Length**. Length$A$ is

$$\sup\{|X| : X \text{ is a subset of } A \text{ totally ordered by the Boolean ordering}\}.$$

6. **Irredundance**. A subset X of a BA A is *irredundant* if for all $x \in X$, $x \notin \langle X \backslash \{x\}\rangle$. The *irredundance* of A, denoted by Irr$A$, is

$$\sup\{|X| : X \text{ is an irredundant subset of } A\}.$$

7. **Cardinality**. This is just $|A|$.

8. **Independence**. A subset X of A is called *independent* if X is a set of free generators for $\langle X\rangle$. Then the *independence* of A, denoted by Ind$A$, is

$$\sup\{|X| : X \text{ is an independent subset of } A\}.$$

9. **π-character**. For any ultrafilter F on A, let $\pi\chi F = \min\{|X| : X$ is dense in $F\}$. Note here that it is not required that $X \subseteq F$. Then the *π-character* of A, denoted by $\pi\chi A$, is

$$\sup\{\pi\chi F : F \text{ an ultrafilter of } A\}.$$

10. **Tightness**. For any ultrafilter F on A, let $\mathrm{t}F = \min\{\kappa :$ if Y is contained in Ult$A$ and F is contained in $\bigcup Y$, then there is a subset Z of Y of power at most κ such that F is contained in $\bigcup Z\}$. Then the *tightness* of A, denoted by $\mathrm{t}A$, is

$$\sup\{\mathrm{t}F : F \text{ is an ultrafilter on } A\}.$$

11. **Spread**. The *spread* of A, denoted by $\mathrm{s}A$, is

$\sup\{|D| : D \subseteq \text{Ult}A$, and D is discrete in the relative topology$\}$.

12. **Character**. The *character* of A, denoted by χA, is

$\min\{\kappa$:every ultrafilter on A can be generated by at most κ elements$\}$.

13. **Hereditary Lindelöf degree**. For any topological space X, the *Lindelöf degree* of X is the smallest cardinal LX such that every open cover of X has a subcover with at most LX elements. Then the *hereditary Lindelöf degree* of A, denoted by hLA, is

$\sup\{\text{L}X : X$ is a subspace of Ult$A\}$.

14. **Hereditary density**. The *hereditary density* of A, hd$A$, is

$\sup\{\text{d}S : S$ is a subspace of Ult$A\}$.

15. **Incomparability**. A subset X of A is *incomparable* if for any two distinct elements $x, y \in X$ we have $x \not\leq y$ and $y \not\leq x$. The *incomparability* of A, denoted by Inc$A$, is

$\sup\{|X| : X$ is an incomparable subset of $A\}$.

16. **Hereditary cofinality**. This cardinal function, h-cof$A$, is

$\min\{\kappa$:for all $X \subseteq A$ there is a $C \subseteq X$ with $|C| \leq \kappa$ and C cofinal in $X\}$.

17. **Number of ultrafilters**. Of course, this is the same as the cardinality of the Stone space of A, and is denoted by $|\text{Ult}A|$.

18. **Number of automorphisms**. We denote by Aut$A$ the set of all automorphisms of A. So this cardinal function is $|\text{Aut}A|$.

19. **Number of endomorphisms**. We denote by End$A$ the set of all endomorphisms of A, and hence this cardinal function is $|\text{End}A|$.

20. **Number of ideals of** A. We denote by IdA the set of all ideals of A, so here we have the cardinal function $|\text{Id}A|$.

21. **Number of subalgebras of** A. We denote by Sub$A$ the set of all subalgebras of A; $|\text{Sub}A|$ is this cardinal function.

Algebraic properties of a single function.

Now we go into more detail on the properties of a single function which we shall investigate. From the point of view of general algebra, the main questions are: what happens to the cardinal function k under the passage to subalgebras, homomorphic images, products, weak products, and free products? In some cases we talk about other operations also: amalgamated free products, unions of well-ordered chains of subalgebras, ultraproducts, dense subalgebras, subdirect products, sheaves of global sections, Boolean products, Boolean powers, set products, one-point gluing, and Alexandroff duplication (these will be explained in the section on cellularity). Also, one may notice that several of the above functions, such as depth and spread, are defined as supremums of the cardinalities of sets satisfying some property P. So, a natural question is whether such sups are *attained*, that is, with depth as an example, whether for every BA A there always is a subset X well-ordered by the Boolean ordering, with $|X| = \mathrm{Depth}A$. Of course, this is only a question in case $\mathrm{Depth}A$ is a limit cardinal. For such functions k defined by sups, we can define a closely related function k'; $k'A$ is the least cardinal such that there is no subset of A with the property P. So $k'A = (kA)^+$ if k is attained, and $k'A = kA$ otherwise.

Derived operations.

From a given cardinal function one can define several others; part of our work is to see what these new cardinal functions look like; frequently it turns out that they coincide with another of our basic 21 functions, but sometimes we arrive at a new function in this way:

$$k_{\mathrm{H}+}A = \sup\{kB : B \text{ is a homomorphic image of } A\}.$$
$$k_{\mathrm{H}-}A = \inf\{kB : B \text{ is an infinite homomorphic image of } A\}.$$
$$k_{\mathrm{S}+}A = \sup\{kB : B \text{ is a subalgebra of } A\}.$$
$$k_{\mathrm{S}-}A = \inf\{kB : B \text{ is an infinite subalgebra of } A\}.$$
$$k_{\mathrm{h}+}A = \sup\{kY : Y \text{ is a subspace of } \mathrm{Ult}A\}.$$
$$k_{\mathrm{h}-}A = \inf\{kY : Y \text{ is an infinite subspace of } \mathrm{Ult}A\}.$$
$${}_{\mathrm{d}}k_{\mathrm{S}+}A = \sup\{kB : B \text{ is a dense subalgebra of } A\}.$$
$${}_{\mathrm{d}}k_{\mathrm{S}-}A = \inf\{kB : B \text{ is a dense subalgebra of } A\}.$$

Note that $k_{\mathrm{h}+}A$ and $k_{\mathrm{h}-}A$ make sense only if k is a function which naturally applies to topological spaces in general as well as BA's. Any infinite Boolean space has a denumerable discrete subspace, and frequently $k_{\mathrm{h}-}$ will take its value on such a subspace (not always—for example, $\mathrm{Length}_{\mathrm{h}-}A$ is always ω, but a denumerable discrete space has length 2^ω).

Given a function defined in terms of ultrafilters, like character above, there is usually an associated function l assigning a cardinal number to each ultrafilter on A. Then one can introduce two cardinal functions on A itself:

$l_{\text{sup}}A = \sup\{lF : F \text{ is an ultrafilter on } A\}$.
$l_{\text{inf}}A = \inf\{lF : F \text{ is a non-principal ultrafilter on } A\}$.

The derived functions so far described are really cardinal functions. For functions which are relatively well-known, there are some more involved notions which are of interest. We consider the following two spectrum functions:

$k_{\text{Hs}}A = \{kB : B \text{ is an infinite homomorphic image of } A\}$ (the ***homomorphic spectrum*** of A).

$k_{\text{Ss}}A = \{kB : B \text{ is an infinite subalgebra of } A\}$ (the ***subalgebra spectrum*** of A).

It is also possible to define a *caliber* notion for many of our functions, in analogy to the well-known caliber notion for cellularity. Given a property P associated with a cardinal function, a BA A is said to have κ, λ,P-*caliber* if among any set of λ elements of A there are κ elements with property P. It is hard to be very precise about this notion in general, but it has been extensively studied for cellularity, and studied somewhat for independence.

Comparing two functions

Given two cardinal functions k and l, one can try to determine whether $kA \leq lA$ for every BA A or $lA \leq kA$ for every BA A. Given that one of these cases arises, it is natural to consider whether the difference can be arbitrarily large (as with cellularity and spread, for example), or if it is subject to restrictions (as with depth and length). If no general relationship is known, a counterexample is needed, and again one can try to find a counterexample with an arbitrarily large difference between the two functions. Of course, the known inequalities between our functions help in order to limit the number of cases that need to be considered for constructing such counterexamples; here the diagrams at the end of these notes are sometimes useful. For example, knowing that $\pi\chi$ can be greater than c, we also know that χ can be greater than c.

Other considerations

In addition to the above systematic goals in discussing cardinal functions, there are some more ideas which we shall not explore in such detail. One can compare several cardinal functions, instead of just two at a time. Several deep theorems of this sort are known, and we shall mention a few of them. There are also a large number of relationships between cardinal functions which involve cardinal arithmetic; for example, $\text{Length}A \leq 2^{\text{Depth}A}$ for any BA A. We mention a few of these as we go along.

One can compare two cardinal functions while considering algebraic operations; for example, comparing functions k, l with respect to the formation of subalgebras. We shall investigate just two of the many possibilities here:

$k_{\mathrm{Sr}}A = \{(\kappa, \lambda)$:there is an infinite subalgebra B of A such that $|B| = \lambda$ and $kB = \kappa\}$.

$k_{\mathrm{Hr}}A = \{(\kappa, \lambda)$:there is an infinite homomorphic image B of A such that $|B| = \lambda$ and $kB = \kappa\}$.

For each function k, it would be nice to be able to characterize the possible relations k_{Sr} and k_{Hr} in purely cardinal number terms.

Special classes of Boolean algebras

We are interested in all of the above ideas not only for the class of all BA's, but also for various important subclasses, like the class of all atomic algebras, all complete algebras, all CSP algebras, all interval algebras, all superatomic algebras, and all atomless algebras.

In closing this section, we should mention that the full treatment of all of the above ideas has only been carried through for one cardinal function, namely cellularity.

1. CELLULARITY

A BA A is said to satisfy the *κ-chain condition* if every disjoint subset of A has power $< \kappa$. Thus for κ non-limit, this is the same as saying that the cellularity of A is $< \kappa$. Of most interest is the ω_1-chain condition, called ccc for short (countable chain condition). We shall return to it below.

The attainment problem for cellularity is covered by two classical theorems of Erdös and Tarski: see Handbook, Part I, Theorem 3.10 and Example 11.14. Cellularity is attained for any singular cardinal, while for weakly inaccessible cardinals there are examples of BA's with cellularity not attained.

Concerning algebraic operations on BA's, for cellularity we shall consider all of the notions mentioned in the introduction.

If B is a subalgebra of A, then obviously $\mathrm{c}B \leq \mathrm{c}A$ and the difference can be arbitrarily large. If B is a dense subalgebra of A, then clearly $\mathrm{c}A = \mathrm{c}B$. If B is a homomorphic image of A, then cellularity can change either way from A to B. For example, if A is a free BA, then it has cellularity ω, while a homomorphic image of A can have very large cellularity. On the other hand, given any infinite BA A, it has a homomorphic image of cellularity ω: take a denumerable subalgebra B of A, and by Sikorski's theorem extend the identity mapping from B into the completion $\overline{B}$ of B to a homomorphism of A into $\overline{B}$.

By an easy argument, $\mathrm{c}(\prod_{i\in I} A_i) = |I| + \sup_{i\in I}\mathrm{c}(A_i)$, if all the A_i are non-trivial. The same computation holds for weak products.

We now describe set products; see Heindorf [87] and Weese [80]. Suppose that $\langle A_i : i \in I\rangle$ is a system of BA's; we assume that A_i is a field of subsets of some set J_i, and that the J_i's are pairwise disjoint. Furthermore, let B be an algebra of subsets of I containing all of the finite subsets of I. For each $b \in B$ let $\overline{b} = \bigcup_{i\in b} J_i$. Set $K = \bigcup_{i\in I} J_i$. For each $b \in B$, each finite $F \subset I$, and each $a \in \prod_{i\in F} A_i$, the set

$$\overline{b} \cup \bigcup_{i\in F} a_i$$

will be denoted by $h(b, F, a)$. It is easily checked that the set of all such elements $h(b, F, a)$ forms a field of subsets of K. This BA is the set product of the A_i's over B, and is denoted by $\prod_{B;i\in I} A_i$. As far as cellularity is concerned they are clearly similar to full direct products:

$$\mathrm{c}\left(\prod_{B;i\in I} A_i\right) = |I| + \sup_{i\in I}\mathrm{c}(A_i).$$

The situation with subdirect products is clear. Suppose that B is a subdirect product of BA's $\langle A_i : i \in I\rangle$; what is the cellularity of B in terms of the cellularity of the A_i's? Well, since a direct product is a special case of a subdirect product, we have the upper bound $cB \leq \sup_{i \in I} cA_i \cup |I|$. The lower bound ω is obvious. And that lower bound can be attained, even if the algebras A_i have high cellularity. In fact, consider the following example. Let κ be any infinite cardinal, let A be the free BA on κ free generators, and let B be the algebra of finite and cofinite subsets of κ. We show that A is isomorphic to a subdirect product of copies of B. To do this, it suffices to take any non-zero element $a \in A$ and find a homomorphism of A onto B which takes $-a$ to 0. In fact, $A \upharpoonright a$ is still free on κ free generators, and so there is a homomorphism of it onto B. So our desired homomorphism is obtained as follows:

$$A \to (A \upharpoonright -a) \times (A \upharpoonright a) \to A \upharpoonright a \to B.$$

Now we turn to sheaves, where we follow the notation of part I of the BA handbook. We recall some notation and results about them. A *sheaf* of BA's is a quadruple $(S, \pi, X, (B_p)_{p \in X})$ satisfying certain conditons; in particular, X is a topological space, and each B_p is a Boolean algebra. We look upon a sheaf as a way of constructing BAs, or decomposing a given BA. The BA in question is the BA GsS of *global sections* of the given sheaf S. It is a subalgebra of the direct product $\prod_{p \in X} B_p$. This gives an upper bound on the cellularity of the BA of global sections. In case X is a Boolean space, this BA is even a subdirect product. Since a direct product is a special case of a BA of global sections, the cellularity of the BA of global sections can attain the maximum possible. Also, it is clear that $cX \leq c(\text{Gs}S)$ when X is a Boolean space; this depends on the following simple fact:

If U is a clopen subset of X and $f \in \prod_{p \in X} B_p$ is defined by $fp = 1$ for $p \in U$ and $fp = 0$ otherwise, then f is continuous.

To prove this, let g and h be the constant functions with values 1 and 0 respectively. Then for any open set V in S we have

$$f^{-1}[V] = (U \cap g^{-1}[V]) \cup ((X \backslash V) \cap h^{-1}[V]),$$

which proves that $f^{-1}[V]$ is open.

Now we describe a construction where the BA of global sections has low cellularity, even though all of the factors B_p have high cellularity.

Let κ be an infinite cardinal (think of it as being large), and let $\lambda = (2^\kappa)^+$. Decompose λ into three disjoint subsets $\Gamma_0, \Gamma_1, \Gamma_2$ each of size λ, and let f_i be a one-one mapping of λ onto Γ_i for each $i < 3$. For convenience we denote $f_i\alpha$ by αi. Now let B be freely generated by free generators $\langle x_\alpha : \alpha < (2^\kappa)^+ \rangle$, and let I be the ideal in B generated by the set

$$\{x_{\alpha 2} \cdot x_{\beta 2} + -x_{\alpha 3} + -x_{\beta 3} : \alpha \neq \beta\}.$$

Let A be the subalgebra $I \cup -I$ of B. Now we apply the standard construction of a sheaf associated with the pair A, B (see the Handbook, Part I, 8.16). By Handbook, Part I, 8.17, B is isomorphic to the BA of global sections of this sheaf. And B has cellularity ω. We now show that each factor has cellularity at least κ^+. Take any factor B_p. Thus $B_p = B/\bar{p}$, where p is an ultrafilter on A and

$$\bar{p} = \{b \in B : a \leq b \text{ for some } a \in p\}.$$

We prefer to work with the ideals $P, \bar{P}$ dual to $p, \bar{p}$. By the Erdös-Rado theorem, we can consider two cases:

Case 1. There is a subset Δ of λ of power κ^+ such that for any two distinct $\alpha, \beta \in \Delta$ we have $x_{\alpha 2} \cdot x_{\beta 2} + -x_{\alpha 3} + -x_{\beta 3} \in P$. For any $\alpha \in \Delta$ let $b_\alpha = [x_{\alpha 1} \cdot x_{\alpha 2}]$; this clearly gives a system of non-zero disjoint elements of $B/\bar{P}$ of size κ^+.

Case 2. There is a subset Δ of λ of power κ^+ such that for any two distinct $\alpha, \beta \in \Delta$ we have $(-x_{\alpha 2} + -x_{\beta 2}) \cdot x_{\alpha 3} \cdot x_{\beta 3} \in P$. For any $\alpha \in \Delta$ let $b_\alpha = [x_{\alpha 1} \cdot -x_{\alpha 2} \cdot x_{\alpha 3}]$. Again this gives the desired disjoint set of size κ^+.

For the notion of *Boolean products* see Burris, Sankappanavar [81], whose notation we follow. We recall the definition. A *Boolean product* of a system $\langle A_x : x \in X \rangle$ of BA's is a subdirect product B of $\langle A_x : x \in X \rangle$ such that X can be endowed with a Boolean topology so that the following conditions hold:

(1) For any two $f, g \in B$ the set $[\![f = g]\!] \stackrel{\text{def}}{=} \{x \in X : fx = gx\}$ is clopen in X.
(2) For any two $f, g \in B$ and any clopen subset N of X, the function $(f \restriction N) \cup (g \restriction (X \backslash N))$ is in B.

(2) is called the *patchwork property.* Boolean products coincide with the global section algebras in which the index space is Boolean and the sheaf space is Hausdorff. To prove this, first suppose that we are given a sheaf $\mathcal{S} = (S, \pi, X, \langle A_x : x \in X \rangle)$ with X Boolean and S Hausdorff. Let B be the algebra of global sections of $\mathcal{S}$. By Theorem 8.13 of the Handbook, part I, B is a subdirect product of $\langle A_x : x \in X \rangle$ and property (1) holds. Assume the hypotheses of (2), and let $h = (f \restriction N) \cup (g \restriction (X \backslash N))$. For any open subset U of S we have $h^{-1}[U] = (N \cap f^{-1}[U]) \cup (g^{-1}[U] \backslash N)$, proving that $h^{-1}[U]$ is open. This checks (2).

The other direction takes more work. Let a Boolean product be given as in the definition above. Without loss of generality we may assume that the sets A_x are pairwise disjoint, and we let $S = \bigcup_{x \in X} A_x$. It is "merely" a matter of putting a topology on S so that we get a sheaf such that B coincides with the BA of global sections of the sheaf. For each $b \in B$ define $f_b : X \to S$ by $f_b x = b_x$. As a base for the desired topology we take

$$\{f_b[U] : U \text{ is clopen in } X, b \in B\}.$$

First we show:

(1) The above set is a base for a topology on S.

To show this, suppose that $a \in f_{b1}[U_1] \cap f_{b2}[U_2]$. Say $a \in A_x$ for a certain $x \in X$. Thus there exist $x_i \in U_i$ such that $a = f_{bi}x_i$ for $i = 1, 2$. Thus $a = (bi)_{x_i}$, and so $x_1 = x_2$. Let

$$V = U_1 \cap U_2 \cap [\![b1 = b2]\!].$$

Then clearly

$$a \in f_{b1}[V] \subseteq f_{b1}[U_1] \cap f_{b2}[U_2],$$

as desired. And for any $s \in S$, say $s \in A_x$, choose $b \in B$ with $b_x = s$. Then $s \in f_b[X]$. This proves (1).

Next, let π be as in Definition 8.14 of Part I of the BA handbook. To show that π is continuous, it suffices to note that for U open in X we have

$$\pi^{-1}[U] = \bigcup_{b \in B} f_b[U].$$

The following statements, easily verified, show that π is an open mapping:

(2) $\pi f_b p = p$.
(3) $\pi[f_b[U]] = U$.

That π is a local homeomorphism also follows easily from (2). Given $s \in S$, say $s \in A_p$ and $b_p = s$, $b \in B$. Then $\pi \restriction f_b[X]$ is one-one by (2), and $f_b[X]$ is a neighborhood of s.

We still need to check the dreaded condition 8.14(d′). We first note that the following simplified form of it implies 8.14(d′) itself in an easy manner:

(4) Let $U \subseteq X$ be open, $f_1, \ldots, f_n$ sections over U, and $1 \leq i \leq n$. Then the set

$$\{p \in U : f_1 p \cdot \cdots \cdot f_i p \cdot -f_{i+1} p \cdot \cdots \cdot -f_n p = 0\}$$

is open.

To prove (4), note that

$$\begin{aligned}
&\{p \in U : f_1 p \cdot \cdots \cdot f_i p \cdot -f_{i+1} p \cdot \cdots \cdot -f_n p = 0\}\\
&= U \cap \bigcup_{b1, \ldots, bn \in B} \{p \in X : f_i p = (bi)p \ (i = 1, \ldots, n) \text{ and}\\
&(b1)p \cdot \cdots \cdot (bi)p \cdot -(b(i+1))p \cdot \cdots \cdot -(bn)p = 0\}\\
&= U \cap \bigcup_{b1, \ldots, bn \in B} \left(\bigcap_{1 \leq i \leq n} \{p \in X : f_i p = (bi)p\} \cap \right.\\
&\left. \{p \in X : [(b1) \cdot \cdots \cdot (bi) \cdot -(b(i+1)) \cdot \cdots \cdot -(bn)]p = 0\} \right);
\end{aligned}$$

hence it suffices to show that each set $\{p \in X : f_i p = (bi)p\}$ is open; but this is clear, since this set is $f_i^{-1}[f_{bi}[X]]$.

Thus we have a sheaf. S is Hausdorff by Theorem 8.13. Now we need to show that B is exactly the set of all global sections with respect to this sheaf. First take any $b \in B$. To show that b is continuous, also take a typical member $f_c[U]$ of the base for the topology on S. Then

$$\begin{aligned} b^{-1}[f_c[U]] &= \{p \in X : b_p = f_c q \text{ for some } q \in U\} \\ &= \{p \in X : b_p = c_q \text{ for some } q \in U\} \\ &= \{p \in X : b_p = c_p \text{ and } p \in U\} \\ &= [\![b = c]\!] \cap U. \end{aligned}$$

On the other hand, suppose that $g \in \prod_{x \in X} A_x$ is continuous; we need to show that $g \in B$. By compactness (as we shall describe in detail below) it suffices to take any $x \in X$ and find a clopen neighborhood U of x and a $b \in B$ such that $g \restriction U = b \restriction U$. In fact, take $b \in B$ such that $b_x = gx$, and let $U = g^{-1}[f_b[X]]$. Then $gx = b_x = f_b x$, so $x \in U$. And if $y \in U$, then

$$\begin{aligned} gy &= f_b z \text{ for some } z \in X \\ &= b_z \text{ for some } z \in X \\ &= b_y, \end{aligned}$$

as desired. We may assume that U is clopen. By compactness, we get a finite sequence $\langle b_i : i < n\rangle$ of elements of B and $\langle U_i : i < n\rangle$ of clopen subsets of X such that $\bigcup_{i<n} U_i = X$ and $g \restriction U_i = b_i \restriction U_i$ for all $i < n$. We may assume that the U_i's are pairwise disjoint. Then the patchwork property of Boolean products yields that $g \in B$.

This completes the proof.

Concerning the cellularity of Boolean products, the situation is like that for sheaves in general. In fact, any BA can be represented as a Boolean product of two element algebras via the Stone isomorphism, so the cellularity can be anything.

Cellularity properties of the related notion of Boolean power reduce to more familiar problems. In fact, bounded Boolean powers coincide with free products, while the unbounded Boolean powers are in between the free product and its completion. (Cellularity of free products is discussed below.) Since these facts are not so well known, we supply a sketch of the proof here (the first fact is due to Quackenbush).

First we recall the definitions of Boolean power and bounded Boolean power, from Burris [75]. Given two BA's A and B, with B complete, the ***Boolean power*** $A[B]$ consists of all $f \in {}^A B$ such that the following two conditions hold:

(1) If $a_0, a_1 \in A$ with $a_0 \neq a_1$ then $fa_0 \cdot fa_1 = 0$;
(2) $\sum_{a \in A} fa = 1$.

The Boolean operations on $A[B]$ are defined like this:

$$(f+g)a = \sum_{b+c=a} (fb \cdot gc);$$

$$(f \cdot g)a = \sum_{b \cdot c=a} (fb \cdot gc);$$

$$(-f)a = f(-a);$$

$$0a = 0, \text{ if } a \neq 0;$$

$$0a = 1, \text{ if } a = 0;$$

$$1a = 0, \text{ if } a \neq 1;$$

$$1a = 1, \text{ if } a = 1.$$

It is easy to verify that $A[B]$ is a BA. The ***bounded Boolean power*** of A and B, denoted by $A[B]^*$, consists of those $f \in A[B]$ such that $\{a \in A : fa \neq 0\}$ is finite; in this case, we do not need to require that B is complete; and clearly $A[B]^*$ is a BA.

Now we define embeddings g of A into $A[B]^*$ and h of B into $A[B]^*$. For any $a \in A$, let

$$(ga)a' = \begin{cases} 1, & \text{if } a = a'; \\ 0, & \text{otherwise.} \end{cases}$$

It is straightforward to check that $ga \in A[B]^*$, and that, in fact, g is an isomorphic embedding of A into $A[B]^*$.

Next, define for any $b \in B$

$$(hb)a = \begin{cases} 0, & \text{if } a \neq 0, 1; \\ b, & \text{if } a = 1; \\ -b, & \text{if } a = 0. \end{cases}$$

Again, it is straightforward to check that $hb \in A[B]^*$, and that h is an isomorphic embedding of B into $A[B]^*$.

If $a \in A^+$ and $b \in B^+$, then $ga \cdot hb \neq 0$. Hence to prove both of the facts mentioned above it suffices to prove the following: for any $\xi \in A[B]$, let $X = \{a \in A^+ : \xi a \neq 0\}$. Then

$$(*) \qquad \xi = \sum_{a \in X} ga \cdot h\xi a.$$

To prove this, it is convenient to prove the following fact first: for $a' \in A^+$ we have

$$(ga \cdot hb)a' = \begin{cases} b, & \text{if } a = a'; \\ 0, & \text{otherwise.} \end{cases}$$

Then it is easy to show that ξ is an upper bound for $\{ga \cdot h\xi a : a \in X\}$. If η is any upper bound for this set, then one can prove the following two facts, valid for any $a \in X$:

$$\xi a = \sum_{a \leq c} \xi a \cdot \eta c;$$

$$\sum_{a \not\leq c} \xi a \cdot \eta c = 0.$$

Then $\xi \leq \eta$ follows easily. Here are the details on these things. Suppose $a \in X$. Then

$$\begin{aligned} \xi a &= (ga \cdot h\xi a)a \\ &= (ga \cdot h\xi a \cdot \eta)a \\ &= \sum_{b \cdot c = a} (ga \cdot h\xi a)b \cdot \eta c \\ &= \sum_{a \cdot c = a} \xi a \cdot \eta c, \end{aligned}$$

giving the first equality above. For the second, if $a \not\leq c$, then

$$\eta c \cdot \xi a = \sum_{a \leq d} \xi a \cdot \eta d \cdot \eta c = 0,$$

which yields the second equality above. Finally, if $a \neq 0$, then

$$\begin{aligned} (\xi \cdot \eta)a &= \sum_{b \cdot c = a} \xi b \cdot \eta c \\ &= \sum_{b \cdot c = a, b \in X} \xi b \cdot \eta c; \end{aligned}$$

if $b \in X$, then $\xi b \cdot \eta c = 0$ unless $b \leq c$, by the second equality above, so this sum is simply $\sum_{a \in X, a \leq c} \xi a \cdot \eta c$, which is ξa by the first equality above. And

$$\begin{aligned} (\xi \cdot \eta)0 &= \sum_{a \cdot b = 0} \xi a \cdot \eta b \\ &= \sum_{a \cdot b = 0, a \in X} \xi a \cdot \eta b + \xi 0 = \xi 0, \end{aligned}$$

as desired.

Our next algebraic operation is one-point gluing. Suppose we are given a system $\langle A_i : i \in I \rangle$ of BA's, and a corresponding system $\langle F_i : i \in I \rangle$ of ultrafilters: F_i is an ultrafilter on A_i for each $i \in I$. The one-point gluing is the following subalgebra of the direct product $\prod_{i\in I} A_i$:

$$B = \{x \in \prod_{i\in I} A_i : \text{for all } i, j \in I(x_i \in F_i \text{ iff } x_j \in F_j)\}.$$

In the case of two factors A_i and A_j this amounts to identifying the two points F_i and F_j in the disjoint union of the Stone spaces. For cellularity, clearly B behaves much like the full direct product: If B is infinite and all algebras A_i have at least four elements, then

$$\mathrm{c}B = |I| + \sup_{i\in I}\mathrm{c}A_i.$$

The behaviour of cellularity under ultraproducts is not completely clear. We mention some obvious results, but we do not have a complete description of what happens. The results that we mention involve the usual notions of *countable completeness*, and *regularity* for ultrafilters. First, if F is a countably complete ultrafilter on an infinite set I and A_i is a ccc BA for each $i \in I$, then $\prod_{i\in I} A_i/F$ also satisfies ccc. This follows from the following standard facts. If F is countably complete and non-principal, then there is an uncountable measurable cardinal, and $|I|$ is at least as big as the first such — call it κ. (See Comfort, Negrepontis [74], p. 196.) Also, F is κ-complete. To see this, suppose not, and let λ be the least cardinal such that F is not λ-complete. Thus $\omega_1 < \lambda \leq \kappa$. Then there exist a cardinal $\mu < \lambda$ and disjoint $a_\alpha \subseteq I$ for $\alpha < \mu$ such that $I\backslash a_\alpha \in F$ for all $\alpha < \mu$, while $\bigcup_{\alpha<\mu} a_\alpha \in F$. Let $G = \{S \subseteq \mu : \bigcup_{\alpha\in S} a_\alpha \notin F\}$. Then it is easy to check that G is a σ-complete non-principal maximal ideal on μ, which is a contradiction, since μ is less than κ.

Now we can give the simple BA argument from these set-theoretical facts. Suppose $\prod_{i\in I} A_i/F$ does not satisfy ccc. Let $\langle [a_\alpha] : \alpha < \omega_1\rangle$ be a system of non-zero disjoint elements of the product; $[x]$ denotes the equivalence class of x under F. Since F is ω_2-complete, the sets $J_{\alpha\beta} \stackrel{\text{def}}{=} \{i \in I : (a_\alpha)_i \cdot (a_\beta)_i = 0\}$ for $\alpha \neq \beta$ and the sets $K_\alpha \stackrel{\text{def}}{=} \{i \in I : (a_\alpha)_i \neq 0\}$ have a non-zero intersection, since that intersection is in F. But this is obviously a contradiction.

Thus countably complete ultrafilters tend to preserve chain conditions; we skip trying to give a more general version of the above argument.

Next, if F is a countably incomplete ultrafilter on I and each algebra A_i is infinite, then $\prod_{i\in I} A_i/F$ never has ccc. This follows from the fact that the product is ω_1-saturated in the model-theoretic sense; see Chang, Keisler [73], p. 305.

Our last result on cellularity in ultraproducts was pointed out by Sabine Koppelberg and Saharon Shelah independently. Namely, let I be infinite, and let F be a $|I|$-regular ultrafilter on I. Suppose that A_i is an infinite BA for each $i \in I$. Then

$$(*)\quad \mathrm{c}\left(\prod_{i\in I} A_i/F\right) \geq \sup\{\min\left\{\prod_{i\in J}\lambda_i : J \in F\right\} : \omega \leq \lambda_i < \mathrm{c}'A_i \text{ for all } i \in I\}.$$

To prove this, assume that $\omega \leq \lambda_i < \mathrm{c}'A_i$ for all $i \in I$. For each $i \in I$, let P_i be a set of λ_i disjoint elements of A_i. Now, using Lemma 2 of Keisler, Prikry [74], we get

$$\begin{aligned}\mathrm{c}\left(\prod_{i\in I} A_i/F\right) &\geq \left|\prod_{i\in I} P_i/F\right| \\ &= \left|\prod_{i\in I}\lambda_i/F\right| \\ &= \left(\min\left\{\sup_{i\in J}\lambda_i : J \in F\right\}\right)^{|I|} \\ &\geq \min\left\{\left(\sup_{i\in J}\lambda_i\right)^{|I|} : J \in F\right\} \\ &\geq \min\left\{\prod_{i\in J}\lambda_i : J \in F\right\},\end{aligned}$$

as desired.

For some further results concerning chain conditions for ultraproducts see Shelah [87].

Now we turn to chain conditions in free products, where there has been a lot of work done. Some partition theorems give results which clarify the situation:

(1) $\mathrm{c}(A \oplus B) \leq 2^{\mathrm{c}A\cdot\mathrm{c}B}$ for infinite BA's A, B.

To see this, suppose that $\langle x_\alpha : \alpha < (2^{\mathrm{c}A\cdot\mathrm{c}B})^+\rangle$ is a system of disjoint elements of $A \oplus B$. Without loss of generality we may assume that for each $\alpha < (2^{\mathrm{c}A\cdot\mathrm{c}B})^+$, the element x_α has the form $a_\alpha \times b_\alpha$, where $a_\alpha \in A$ and $b_\alpha \in B$ (we use $\times$ to make clear that the indicated product of elements is in the algebra $A \oplus B$). Thus for distinct α, β we have $a_\alpha \cdot a_\beta = 0$ or $b_\alpha \cdot b_\beta = 0$, and hence the Erdös-Rado partition theorem $(2^\kappa)^+ \to (\kappa^+)^2_\kappa$ implies that there is a subset Y of $(2^{\mathrm{c}A\cdot\mathrm{c}B})^+$ of power $(\mathrm{c}A \cdot \mathrm{c}B)^+$ such that either $a_\alpha \cdot a_\beta = 0$ for all distinct $\alpha, \beta \in Y$ or $b_\alpha \cdot b_\beta = 0$ for all distinct $\alpha, \beta \in Y$, which is impossible. (For this partition relation, see Erdös, Hajnal, Máté, Rado [84], pp. 98-100.)

Similarly, the partition theorem $(2^\kappa)^+ \to ((2^\kappa)^+, \kappa^+)^2$ (see the above book, Corollary 17.5) gives the following result:

(2) If $\mathrm{c}A \leq 2^\kappa$ and $\mathrm{c}B \leq \kappa$, then $\mathrm{c}(A \oplus B) \leq 2^\kappa$.

Furthermore, if κ is strong limit, then $\kappa^+ \to (\kappa^+, \mathrm{cf}\kappa)$ (see the above book, Theorem 17.1). Hence

(3) If κ is strong limit, $\mathrm{c}A \le \kappa$, and $\mathrm{c}B < \mathrm{cf}\kappa$, then $\mathrm{c}(A \oplus B) \le \kappa$.

The results (1) and (2) were first proved by Kurepa [62].

Under GCH, these results say the following: (1) For any BA's A and B, $\mathrm{c}A \cdot \mathrm{c}B \le \mathrm{c}(A \oplus B) \le (\mathrm{c}A \cdot \mathrm{c}B)^+$; (2) For any BA's A and B, if $\mathrm{c}B < \mathrm{cf}(\mathrm{c}A)$, then $\mathrm{c}(A \oplus B) = \mathrm{c}A$. Thus even under GCH, there are two cases not covered by (1)-(3): when $\mathrm{c}A$ is limit with $\mathrm{cf}(\mathrm{c}A) \le \mathrm{c}B < \mathrm{c}A$, and when $\mathrm{c}A = \mathrm{c}B$. We have no information on the first case:

Problem 1. *If κ is strong limit singular, $\mathrm{c}A = \kappa$, and $\mathrm{cf}\kappa \le \mathrm{c}B < \kappa$, is $\mathrm{c}(A \oplus B) = \kappa$?*

The case $\mathrm{c}A = \mathrm{c}B$ has been intensively studied in the literature. That it is consistent to have a BA A such that $\mathrm{c}(A \oplus A) > \mathrm{c}A$ was essentially recognized quite early (probably at least implicitly by Kurepa); we give such an example shortly. Laver made a major advance by showing that CH suffices for such an example. The first example of such a phenomemon purely in ZFC was given by Todorčević. For the most recent results, see Shelah [86], [88a], [88b], [88c], [89], and Todorčević [85], [86]. The problem in the above form is now completely resolved: For each cardinal $\kappa > \omega$ there is a BA A with $\mathrm{c}A = \kappa$ while $\mathrm{c}(A \oplus A) > \kappa$. Moreover, it is well-known that for ccc the question is independent of ZFC (see below). In the above papers one can still find some interesting open problems, but we shall not state them here; they concern productivity of the κ-chain condition for inaccessible κ (this is different from the cellularity questions because of the possible non-attainment of cellularity).

Perhaps the easiest result concerning this circle of questions is as follows. Let T be a Souslin tree such that every element t has at infinitely many successors, denote two of them by $t0$ and $t1$, and let A be the tree algebra on T. Recall that for any $s \in T$ we denote by T_s the generator $\{u \in T : s \le u\}$ of A. Now A satisfies ccc, but $A \oplus A$ does not. To see this second fact, for each $t \in T$ consider the element $T_{t0} \times T_{t1}$ of $A \oplus A$. Suppose that $s, t \in T$, $s \neq t$, and $T_{t0} \cap T_{s0} \neq 0$. Then $t0$ and $s0$ are comparable; say $s0 < t0$. Clearly, then, $t1$ and $s1$ are not comparable, so $T_{t1} \cap T_{s1} = 0$, as desired.

Another important and quite elementary fact about free products is that

$$\mathrm{c}(\oplus_{i \in I} A_i) = \sup\{c(\oplus_{i \in F} A_i) : F \in [I]^{<\omega}\}.$$

In fact, $\ge$ is clear. Now let $\kappa = \sup\{\mathrm{c}(\oplus_{i \in F} A_i) : F \in [I]^{<\omega}\}$, and suppose that X is a disjoint subset of $\oplus_{i \in I} A_i$ of size κ^+. For each $x \in X$ choose a finite $Fx \subseteq I$ such that $x \in \oplus_{i \in Fx} A_i$. We may assume that each $x \in X$ has the form $x = \prod_{i \in Fx} y_i^x$, where $y_i^x \in A_i$ for each $i \in Fx$. Without loss of generality, $\langle Fx : x \in X\rangle$ forms a Δ-system, say with kernel G. But then, by the free product property, $\langle\prod_{i\in G} y_i^x : x \in X\rangle$ is a disjoint system of elements of $\oplus_{i\in G}A_i$, contradiction.

As our final result on chain conditions in free products, we prove the folklore theorem that MA (Martin's axiom) + $\neg$CH implies that the free product of two ccc BA's is again ccc. This depends on the following lemma:

Lemma. (MA + $\neg$CH) *Suppose that $\langle x_\alpha : \alpha < \omega_1\rangle$ is a system of elements in a ccc BA A. Then there is an uncountable $S \subseteq \omega_1$ such that $\langle x_\alpha : \alpha \in S\rangle$ has the finite intersection property.*

PROOF. We may assume that A is complete. For each $\alpha < \omega_1$ let $y_\alpha = \sum_{\gamma>\alpha} x_\gamma$. Then, we claim,

(*) There is an $\alpha < \omega_1$ such that for all $\beta > \alpha$ we have $y_\beta = y_\alpha$.

Otherwise, since clearly $\alpha < \beta \rightarrow y_\alpha \geq y_\beta$, we easily get an increasing sequence $\langle\beta(\xi) : \xi < \omega_1\rangle$ of ordinals less than ω_1 such that $y_{\beta(\xi)} > y_{\beta(\eta)}$ whenever $\xi < \eta < \omega_1$. But then $\langle y_{\beta(\xi)} \cdot -y_{\beta(\xi+1)}\rangle$ is a disjoint family of power ω_1, contradiction.

Thus (*) holds, and we fix an α as indicated there. The partial ordering P that we want to apply Martin's axiom to is $\{x \in A : 0 \neq x \leq y_\alpha\}$ under $\geq$. It is a ccc partial ordering since A is a ccc BA. Now for the dense sets. For each $\beta < \omega_1$ let

$$D_\beta = \{p \in P : \text{there is a } \gamma > \beta \text{ such that } p \leq x_\gamma\}.$$

To see that D_β is dense in P, let $p \in P$ be arbitrary. Choose $\delta \in \omega_1$ with $\delta > \alpha, \beta$. Then $y_\alpha = y_\delta$, so from $0 \neq p \leq y_\alpha$ we infer that there is a $\gamma > \delta$ such that $p \cdot x_\gamma \neq 0$. Thus $p \cdot x_\gamma$ is the desired element of D_β which is $\leq p$.

Now let G be a filter on P intersecting each dense set D_β for $\beta < \omega_1$, by MA. Then it is easy to see that $S \overset{\text{def}}{=} \{x_\gamma : \gamma < \omega_1, \text{ and } p \leq x_\gamma \text{ for some } p \in G\}$ is the set desired in the lemma. □

Now we prove, using MA+$\neg$CH, that the free product of ccc BA's A and B is again ccc. Let $\langle x_\alpha : \alpha < \omega_1\rangle$ be a disjoint system of elements of $A \oplus B$. Without loss of generality we may assume that each x_α has the form $a_\alpha \times b_\alpha$ where $a_\alpha \in A$ and $b_\alpha \in B$. By the lemma, let S be an uncountable subset of ω_1 such that $\langle a_\alpha : \alpha \in S\rangle$ has the finite intersection property. But then, by the free product property, $\langle b_\alpha : \alpha \in S\rangle$ is a disjoint system in B, contradiction.

The argument just given generalizes easily to show that MA+$\neg$CH implies that if X and Y are ccc topological spaces, then so is $X \times Y$.

Next, we consider what happens to cellularity under amalgamated free products. The basic fact is as follows:

$$\mathrm{c}(A \oplus_C B) \leq 2^{\mathrm{c}A\cdot\mathrm{c}B\cdot|C|}.$$

To prove this, let $\kappa = \mathrm{c}A \cdot \mathrm{c}B \cdot |C|$, and suppose that $\langle c_\alpha : \alpha < (2^\kappa)^+ \rangle$ is a disjoint system in $A \oplus_C B$. We may assume that each c_α is non-zero, and has the form $a_\alpha \cdot b_\alpha$, with $a_\alpha \in A$ and $b_\alpha \in B$. Thus for all distinct $\alpha, \beta < (2^\kappa)^+$ there is a $c \in C$ such that $a_\alpha \cdot a_\beta \leq c$ and $b_\alpha \cdot b_\beta \leq -c$. Hence by the Erdös-Rado theorem there is a $\Gamma \in [(2^\kappa)^+]^{\kappa^+}$ and a $c \in C$ such that $a_\alpha \cdot a_\beta \leq c$ and $b_\alpha \cdot b_\beta \leq -c$ for all distinct $\alpha, \beta \in \Gamma$. Thus $(a_\alpha \cdot -c) \cdot (a_\beta \cdot -c) = 0$ and $(b_\alpha \cdot c) \cdot (b_\beta \cdot c) = 0$ for all distinct $\alpha, \beta \in (2^\kappa)^+$. Since $\mathrm{c}A < \kappa^+$, it follows that there is a $\Delta \in [\Gamma]^\kappa$ such that $a_\alpha \cdot -c = 0$ for all $\alpha \in \Gamma \backslash \Delta$; and there is a $\Theta \in [\Gamma \backslash \Delta]^\kappa$ such that $b_\alpha \cdot c = 0$ for all $\alpha \in (\Gamma \backslash \Delta) \backslash \Theta$. But then for any $\alpha \in (\Gamma \backslash \Delta) \backslash \Theta$ we have $a_\alpha \cdot b_\alpha = 0$, contradiction.

The above inequality is best-possible, in a sense. To see this, consider $\mathcal{P}\omega \oplus_C \mathcal{P}\omega$, where C is the BA of finite and cofinite subsets of ω. Let $\langle \Gamma_\alpha : \alpha < 2^\omega \rangle$ be a system of infinite almost disjoint subsets of ω; and also assume that each Γ_α is not cofinite. For each $\alpha < 2^\omega$ let y_α be the element $\Gamma_\alpha \cdot (\omega \backslash \Gamma_\alpha)$ of $\mathcal{P}\omega \oplus_C \mathcal{P}\omega$. These elements are clearly non-zero. For distinct $\alpha, \beta < 2^\omega$ let $F = \Gamma_\alpha \cap \Gamma_\beta$. Then $\Gamma_\alpha \cap \Gamma_\beta = F$ and $(\omega \backslash \Gamma_\alpha) \cap (\omega \backslash \Gamma_\beta) \subseteq (\omega \backslash F)$, which shows that the system is disjoint. This demonstrates equality above.

For free amalgamated products with infinitely many factors we have

$$(A) \qquad \mathrm{c}(\oplus^C_{i \in I} A_i) \leq 2^{|C|} \cdot 2^{\sup_{i \in I} \mathrm{c}A_i}.$$

To prove this, let κ be the cardinal on the right, and suppose that $\langle y_\alpha : \alpha < \kappa^+ \rangle$ is a disjoint system of elements of $\oplus^C_{i \in I} A_i$. We may assume that each y_α has the form

$$y_\alpha = \prod_{i \in F_\alpha} a_i^\alpha,$$

where F_α is a finite subset of I and $a_i^\alpha \in A_i$ for all $i \in F_\alpha$. We may assume, in fact, that the F_α's form a Δ-system, say with kernel G; and that they all have the same size. Thus by a change of notation we may write

$$y_\alpha = \prod_{j<m} a^\alpha_{i^\alpha_j} \cdot \prod_{j<n} a^\alpha_{k_j},$$

where $F_\alpha \backslash G = \{i^\alpha_j : j < m\}$ and $G = \{k_j : j < n\}$. For distinct $\alpha, \beta < \kappa^+$ there then exist $c_j \in C$ for $j < m$, $d_j \in C$ for $j < m$, and $e_j \in C$ for $j < n$ such that $a^\alpha_{i^\alpha_j} \leq c_j$ for all $j < m$, $a^\beta_{i^\beta_j} \leq d_j$ for all $j < m$, and $a^\alpha_{k_j} \cdot a^\beta_{k_j} \leq e_j$ for all $j < n$, such that $\prod_{j<m} c_j \cdot \prod_{j<m} d_j \cdot \prod_{j<n} e_j = 0$. Using the Erdös-Rado theorem again, we get $\Gamma \in [\kappa^+]^\lambda$ and $c, d \in {}^mC$, $e \in {}^nC$ such that the above holds for all distinct $\alpha, \beta \in \Gamma$, where $\lambda = (|C| \cdot \sup_{i \in I} \mathrm{c}A_i)^+$. Arguing similarly to the case of a free product with amalgamation of two algebras, we then easily infer that there is an $\alpha \in \Gamma$ such that $a^\alpha_{k_j} \leq e_j$ for each $j < n$. But then $y_\alpha = 0$, contradiction.

Since $\mathcal{P}\omega \oplus_C \mathcal{P}\omega$ can be considered as a subalgebra of $\oplus^C_{i\in\omega}\mathcal{P}\omega$, with C as in the example for the free product of two factors, it follows that the inequality (A) is again best possible.

The behaviour of cellularity under unions of well-ordered chains is clear on the basis of cardinal arithmetic. We restrict ourselves, without loss of generality, to well-ordered chains of regular type. Actually, we can formulate a more general fact about increasing chains of BA's; this fact will apply to several of our cardinal functions.

A cardinal function k is an *ordinary sup-function* with respect to P if P is a function assigning to every infinite BA A a subset PA of $\mathcal{P}A$ so that the following conditions hold for any infinite BA A:

(1) $kA = \sup\{|X| : X \in PA\}$;

(2) If B is a subalgebra of A, then $PB \subseteq PA$ and $X \cap B \in PB$ for any $X \in PA$.

(3) For each infinite cardinal κ there is a BA C of size κ such that there is an $X \in PC$ with $|X| = \kappa$.

Note that cellularity is an ordinary sup-function with respect to the following fuction P: for any infinite BA A, $PA = \{X \subseteq A : X \text{ is disjoint}\}$. Given any ordinary sup-function k with respect to a function P and any infinite cardinal κ, we say that A satisfies the $\kappa - k-$*chain condition* provided that $|X| < \kappa$ for all $X \in PA$. This agrees with the usual notion for cellularity; see the beginning of this section.

Theorem 1.1. *Let κ and λ be infinite cardinals, with λ regular. Suppose that k is an ordinary sup-function with respect to P. Then the following conditions are equivalent:*

(i) $\mathrm{cf}\kappa = \lambda$.

(ii) There is a strictly increasing sequence $\langle A_\alpha : \alpha < \lambda\rangle$ of BA's each satisfying the $\kappa - k-$chain condition such that $\bigcup_{\alpha<\lambda} A_\alpha$ does not satisfy this condition.

Proof. (i)⇒(ii): Assume (i). Let $\langle \mu_\xi : \xi < \lambda\rangle$ be a strictly increasing sequence of *ordinals* with sup κ (maybe κ is a successor cardinal, so that we cannot take the μ_ξ to be *cardinals*). Let A be a BA of size κ with a set $X \in PA$ such that $|X| = \kappa$. Write $A = \{a_\alpha : \alpha < \kappa\}$. For each $\xi < \lambda$ let $B_\xi = \langle\{a_\alpha : \alpha < \mu_\xi\}\rangle$. Thus $B_\xi \subseteq B_\eta$ if $\xi < \eta$, and $|B_\xi| < \kappa$ for all $\xi < \lambda$. Hence a strictly increasing subsequence is as desired (since λ is regular).

(ii)⇒(i). Assume that (ii) holds but (i) fails. Let X be a subset of $\bigcup_{\alpha<\lambda} A_\alpha$ of power κ which is in PA. If $\lambda < \mathrm{cf}\kappa$, then the facts that $X = \bigcup_{\alpha<\lambda}(X \cap A_\alpha)$, $|X| = \kappa$, and $|X \cap A_\alpha| < \kappa$ for all $\alpha < \lambda$, give a contradiction.

So, assume that $\mathrm{cf}\kappa < \lambda$. Now for all $\alpha < \lambda$ there is a $\beta > \alpha$ such that $X \cap A_\alpha \subset X \cap A_\beta$, since otherwise some A_α would contain X. It follows that $\lambda \leq \kappa$, and so κ is singular in the case we are considering. Let $\langle \mu_\alpha : \alpha < \mathrm{cf}\kappa\rangle$ be a strictly increasing sequence of cardinals with sup κ. Since $\sup_{\alpha<\lambda}|X \cap A_\alpha| = \kappa$, for each $\alpha < \mathrm{cf}\kappa$, choose $\nu(\alpha) < \lambda$ such that $|X \cap A_{\nu(\alpha)}| \geq \mu_\alpha$. Let $\rho = \sup_{\alpha<\mathrm{cf}\kappa}\nu(\alpha)$. Then $\rho < \lambda$ since $\mathrm{cf}\kappa < \lambda$ and λ is regular. But $|X \cap A_\rho| = \kappa$, contradiction. □

With regard to Theorem 1.1, see also the end of this section.

Our last algebraic operation is Alexandroff duplication. Given a BA A, its *Alexandroff duplicate*, denoted by $\mathrm{Dup}A$, is the subalgebra of $A \times \mathcal{P}\mathrm{Ult}A$ whose set of elements is

$$\{(a, X) : a \in A, X \subseteq \mathrm{Ult}A, \text{ and } \mathcal{S}a \triangle X \text{ is finite}\}.$$

(It is easy to check that this is a subalgebra of $A \times \mathcal{P}\mathrm{Ult}A$; recall that $\mathcal{S}a = \{F \in \mathrm{Ult}A : a \in F\}$.) Clearly $\mathrm{c}(\mathrm{Dup}A) = |\mathrm{Ult}A|$.

Now we proceed to discuss the derived functions associated with cellularity. First we show that $\mathrm{c}_{\mathrm{H}+}$ is the same as spread. For this, it is convenient to have an equivalent definition of spread. A subset X of a BA A is *ideal independent* if $x \notin \langle X\backslash\{x\}\rangle^{\mathrm{Id}}$ for every $x \in X$; recall that $\langle Y\rangle^{\mathrm{Id}}$ denotes the ideal generated by Y, for any $Y \subseteq A$.

Theorem 1.2. *For any infinite BA A, $\mathrm{s}A = \sup\{|X| : X$ is an ideal independent subset of $A\}$.*

PROOF. First suppose that D is a discrete subspace of $\mathrm{Ult}A$. For each $F \in D$, let $a_F \in A$ be such that $\mathcal{S}a_F \cap D = \{F\}$. Then $\langle a_F : F \in D\rangle$ is one-one and $\{a_F : F \in D\}$ is ideal independent. In fact, suppose that $F, G_0, \ldots, G_{n-1}$ are distinct members of D such that $a_F \leq a_{G_0} + \ldots + a_{G_{n-1}}$. Then $\mathcal{S}a_F \subseteq \mathcal{S}a_{G_0} \cup \ldots \cup \mathcal{S}a_{G_{n-1}}$, and so $F \in \mathcal{S}a_{G_0} \cup \ldots \cup \mathcal{S}a_{G_{n-1}}$, which is clearly impossible.

Conversely, suppose that X is an ideal independent subset of A. Then for each $x \in X$, $\{x\} \cup \{-y : y \in X\backslash\{x\}\}$ has the finite intersection property, and so is included in an ultrafilter F_x. Let $D = \{F_x : x \in X\}$. Then $\mathcal{S}x \cap D = \{F_x\}$ for each $x \in X$, so D is discrete and $|D| = |X|$, as desired. □

Theorem 1.3. *For any infinite BA A, $\mathrm{c}_{\mathrm{H}+}A$ is equal to $\mathrm{s}A$, the spread of A.*

PROOF. First let f be a homomorphism from A onto a BA B, and let X be a disjoint subset of B. We show that $|X| \leq \mathrm{s}A$; this will show that $\mathrm{c}_{\mathrm{H}+}A \leq \mathrm{s}A$. For each $x \in X$ choose $a_x \in X$ such that $fa_x = x$. Then $\langle a_x : x \in X\rangle$ is one-one and $\{a_x : x \in X\}$ is ideal independent. In fact, suppose that $x, y(0), \ldots y(n-1)$ are distinct elements of X, and $a_x \leq a_{y(0)} + \cdots + a_{y(n-1)}$.

Applying the homomorphism f to this inequality we get $x \leq y(0)+\cdots+y(n-1)$. Since the elements $x, y(0), \ldots, y(n-1)$ are disjoint, this is impossible.

For the converse, suppose that X is an ideal independent subset of A; we want to find a homomorphic image B of A having a disjoint subset of size $|X|$. Let $I = \langle\{x \cdot y : x, y \in X, x \neq y\}\rangle^{\mathrm{Id}}$. It suffices now to show that $[x] \neq 0$ for each $x \in X$. ($[u]$ is the equivalence class of u under the equivalence relation naturally associated with the ideal I). Suppose that $[x] = 0$. Then x is in the ideal I, and hence there exist elements $y_0, z_0, \ldots, y_{n-1}, z_{n-1}$ of X such that $y_i \neq z_i$ for all $i < n$, and $x \leq y_0 \cdot z_0 + \ldots + y_{n-1} \cdot z_{n-1}$. Without loss of generality, $x \neq y_i$ for all $i < n$. But then $x \leq y_0 + \ldots y_{n-1}$, contradicting the ideal independence of X. □

For later purposes it is convenient to note the following corollary to the proof of the previous two theorems.

Corollary 1.4. $\mathrm{c_{H+}}A$ *and* $\mathrm{s}A$ *have the same attainment properties, in the sense that* $\mathrm{s}A$ *is attained (in either the discrete subspace or ideal independence sense) iff there exist a homomorphic image* B *of* A *and a disjoint subset* X *of* B *such that* $|X| = \mathrm{c_{H+}}A$. □

Note in this corollary that attainment of $\mathrm{c_{H+}}A$ involves two sups, while attainment of $\mathrm{s}A$ involves only one. Thus if $\mathrm{s}A$ is not attained, there are still two possibilities according to Corollary 1.4: there can exist a homomorphic image B of A with $\mathrm{s}A = \mathrm{c}B$ but $\mathrm{c}B$ is not attained, or there is no homomorphic image B of A with $\mathrm{s}A = \mathrm{c}B$. Both possiblities are consistent with ZFC; we shall return to this shortly and indicate the examples.

It is easy to see that $\mathrm{c_{H-}}A = \omega$ for any infinite BA A: let B be a denumerable subalgebra of A, and extend the identity homomorphism h of B into $\overline{B}$ to a homomorphism from A into $\overline{B}$; the image of A under h is a ccc BA. (We are using here Sikorski's extension theorem; recall that $\overline{B}$ is the completion of B.) It is obvious that $\mathrm{c_{S+}}A = \mathrm{c}A$ and $\mathrm{c_{S-}}A = \omega$ for any infinite BA A. $\mathrm{c_{h+}}A$ is equal to $\mathrm{s}A$, since a disjoint family of open subsets of a subspace Y of $\mathrm{Ult}A$ gives a discrete subset of $\mathrm{Ult}A$ of the same size, so that $\mathrm{c_{h+}}A \leq \mathrm{s}A = \mathrm{c_{H+}}A \leq \mathrm{c_{h+}}A$. It is obvious that $\mathrm{c_{h-}}A = \omega$, and an easy argument gives that ${}_{\mathrm{d}}\mathrm{c_{S+}}A = \mathrm{c}A = {}_{\mathrm{d}}\mathrm{c_{S-}}A$.

The homomorphic spectrum of cellularity is more interesting. First, we can easily see that $[\omega, \mathrm{s}A) \subseteq \mathrm{c_{Hs}}A \subseteq [\omega, \mathrm{s}A]$ (for cardinals $\kappa < \lambda$, $[\kappa, \lambda)$ denotes the set of all cardinals μ such that $\kappa \leq \mu < \lambda$; similarly for $[\kappa, \lambda]$). This follows from the fact already proved that $\mathrm{s}A = \mathrm{c_{H+}}A$: given a homomorphic image B of A and a disjoint subset X of B, one can use Sikorski's extension theorem to get a homomorphic image C of B such that $\mathrm{c}C = |X|$.

It is more difficult to decide whether $\mathrm{s}A \in \mathrm{c_{Hs}}A$. This amounts to the following question: is there always a homomorphic image B of A such that $\mathrm{c}B = \mathrm{s}A$? In case $\mathrm{s}A$ is attained, this is true by Corollary 1.4. For $\mathrm{s}A$ not attained, there are three consistency results which clarify things here and with respect to the question raised above after Corollary 1.4. First, example 11.14 in Part I of the BA handbook shows that for each weakly inaccessible cardinal κ there is a BA A such that $|A| = cA = sA = \kappa$ and cA is not attained but sA is attained (as is easily checked). Second, the interval algebra of a κ-Souslin line, for κ strongly inaccessible but not weakly compact, gives an example of a BA A such that $|A| = cA = sA$, with neither cA nor sA attained; see Juhász [71], example 6.6 (V=L is needed for the existence of a κ-Suslin line). Third, an example of Todorčević [86], Theorem 12, shows that it is consistent to have a BA A in which sA is not attained, while there is no homomorphic image B of A with $cB{=}sA$. This example involves some interesting ideas, and we shall now give it. It depends on the following lemma about the real numbers.

Lemma 1.5. *There exist disjoint subsets E_0 and E_1 of $[0,1]$ which are of cardinality 2^ω, are dense in $[0,1]$, and satisfy the following two conditions:*

(i) For any $\kappa < 2^\omega$ there is a strictly increasing function from some subset of E_0 of size κ into E_1.

(ii) There is no strictly monotone function from a subset of E_0 of size 2^ω into E_1.

PROOF. The idea of the proof is to construct E_0 and E_1 in steps, "killing" all of the possible big strictly monotone functions as we go along. The very first thing to do is to see that we can list out in a sequence of length 2^ω all of the functions to be "killed".

For the empty set 0 we let sup0=0, inf0=1. For any subset W of [0,1] we let cl$W$ be its topological closure in [0,1], and we let $C_1W = \{f : f : W \to [0,1]$, and f is either strictly increasing or strictly decreasing$\}$. For $W \subseteq [0,1]$ and $f \in C_1W$ (say f strictly increasing) we define $f_{cl} : \text{cl}W \to [0,1]$ by

$$f_{cl}x = \begin{cases} fx \text{ if } x \in W, \\ \sup\{fy : x > y \in W\} \text{ if } x \notin W \text{ and } x = \sup\{y \in W : y < x\}, \\ \inf\{fy : x < y \in W\} \text{ if } x \notin W \text{ and } x \neq \sup\{y \in W : y < x\}. \end{cases}$$

(A similar definition is given if f is strictly decreasing.) Note that if $x \in \text{cl}W\backslash W$ then $x = \sup\{y \in W : y < x\}$ or $x = \inf\{y \in W : x < y\}$. Hence if $x, y \in \text{cl}W\backslash W$, $x < y$, and $f_{cl}x = f_{cl}y$, then $x = \sup\{z \in W : z < x\}$, $y = \inf\{z \in W : y < z\}$, and $|(x,y) \cap W| \leq 1$. Hence for each $z \in [0,1]$, the set $f_{cl}^{-1}[\{z\}]$ has at most three elements. Furthermore, f strictly increasing (strictly decreasing) implies that f_{cl} is increasing (decreasing).

For any $W \subseteq [0,1]$ let C_2W be the set of all functions $f : W \to I$ such that

(1) f is either increasing or decreasing,

(2) $f^{-1}[\{y\}]$ is finite for all $y \in [0,1]$,

(3) $|\{x \in W : fx \neq x\}| = 2^\omega$.

Thus by the above, $f_{\mathrm{cl}} \in C_2W$ whenever $f \in C_1W$ and $|W| = 2^\omega$.

Given $f \in C_2W$, say f increasing, let F_1 be a countable subset of W which is dense in W in the sense that if $x, y \in W$ and $x < y$, then there is a $z \in F_1$ such that $x \leq z \leq y$. Furthermore, let

$$F_2 = \{x \in W : \sup\{fy : x \geq y \in F_1\} < \inf\{fy : x \leq y \in F_1\}\}.$$

If $x \in F_2$, then $x \notin F_1$. Hence if $x, y \in F_2$ and $x < y$, then there is a $z \in F_1$ with $x < z < y$. It follows that the sup and inf above determine an open interval U_x in $\mathbf{R}$ so that $U_x \cap U_y = 0$ for $x \neq y$. So F_2 is countable. Note that f is determined by its restriction to $F_1 \cup F_2$. From these considerations it follows that $|C_2W| \leq 2^\omega$. Also recall that there are just 2^ω closed sets, since every closed set is the closure of a countable dense subset. Hence the set

$$C = \bigcup\{C_2F : F \subseteq [0,1], F \text{ closed}\}$$

has cardinality $\leq 2^\omega$. Let $\langle f_\alpha : \alpha < 2^\omega\rangle$ be an enumeration of C. Let h be a strictly decreasing function from $\mathbf{R}$ onto (0,1); thus h^{-1} is also strictly decreasing. (Here $\mathbf{R}$ is the set of all real numbers.) Moreover, fix a well-ordering of $\mathbf{R}$.

Now we construct by induction pairwise disjoint subsets A_α of $[0,1]$ for $\alpha < 2^\omega$. At the end we will let E_0 be the union of the A_α with even α and E_1 be the union of the rest. We will carry along the inductive hypothesis that $|\alpha| \leq |A_\alpha| \leq |\alpha| + \omega$. Let A_0 and A_1 be denumerable disjoint subsets of [0,1] which are dense in [0,1].

Now suppose that A_α has been constructed for all $\alpha < \beta$, where $\beta \geq 2$. Let $B_\beta = \bigcup_{\alpha<\beta} A_\alpha$ and $B^*_\beta = B_\beta \cup \bigcup_{\alpha<\beta} f_\alpha[B_\beta] \cup \bigcup_{\alpha<\beta} f^{-1}[B_\beta]$. Note by our assumptions that $|\beta| \leq |B^*_\beta| \leq |\beta| + \omega$. For every real number r, let

$$C_r = h[\{r + h^{-1}b : b \in B^*_\beta\}].$$

We claim that there is an $r \in \mathbf{R}$ such that $C_r \cap B^*_\beta = 0$. Suppose not. For every $r \in \mathbf{R}$ choose $b_r \in C_r \cap B^*_\beta$. Since $|B^*_\beta| < 2^\omega$, there exist a set $S \subseteq \mathbf{R}$ and an element $c \in B^*_\beta$ such that $|S| > B^*_\beta$ and $b_r = c$ for all $r \in S$. Say $c = hd$ with $d \in \{r + h^{-1}x : x \in B^*_\beta\}$ for all $r \in S$. Thus $h(d - r) \in B^*_\beta$ for all $r \in S$. So there exist distinct $r, s \in S$ such that $h(d - r) = h(d - s)$. This contradicts h being one-one.

Finally, let r be the least real number (in the well-ordering fixed above) such that $C_r \cap B^*_\beta = 0$, and let $A_\beta = C_r$. Clearly the inductive hypothesis remains true. This finishes the construction of the sets A_α, $\alpha < 2^\omega$.

Let $E_0 = \bigcup_{\alpha \text{ even}} A_\alpha$ and $E_1 = \bigcup_{\alpha \text{ odd}} A_\alpha$. So E_0 and E_1 are disjoint subsets of [0,1], and both of them are dense in [0,1]. Since all of the sets A_α are non-empty, it is clear that both E_0 and E_1 are of power 2^ω.

Now suppose that f is a strictly monotone function from a subset of E_0 of power 2^ω into E_1. Say $f_{\rm cl} = f_\alpha$. Now $|\bigcup_{\beta \leq \alpha} A_\alpha| < 2^\omega$, so choose $y \in \operatorname{ran} f$ such that $y \in A_\gamma$ for some $\gamma > \alpha$. Say $fx = y$ with $x \in A_\delta$. Now δ is even and γ is odd. If $\delta < \gamma$, then $y \in B_\gamma^*$, so $y \in A_\gamma$ is a contradiction. If $\gamma < \delta$, then $x \in B_\delta^*$, so $x \in A_\delta$ is a contradiction. Thus (ii) of the lemma has been verified.

If $\omega \leq \kappa < 2^\omega$, choose $\beta < 2^\omega$ odd with $\beta > \kappa$. Say $A = C_r$, as in the definition. Now $b \mapsto h(r + h^{-1}b)$ is an increasing mapping from B_β^* into A_β, and $E_0 \cap B_\beta^*$ has at least κ elements. This verifies (i). □

The theorem also depends upon the following lemma, which will also be useful later on.

Lemma 1.6. *Let A be the interval algebra of* $\mathbf{R}$. *Then there does not exist in A a strictly increasing sequence $\langle I_\alpha : \alpha < \omega_1 \rangle$ of ideals.*

PROOF. Suppose that there is such a sequence. For each $\alpha < \omega_1$ define $r \equiv_\alpha s$ iff $r, s \in \mathbf{R}$ and either $r = s$ or else if, say, $r < s$, then $[r, s) \in I_\alpha$. Then $\equiv_\alpha$ is an equivalence relation on $\mathbf{R}$ and the equivalence classes are intervals. For each $r \in \mathbf{R}$ the left endpoints of the intervals $[r]_\alpha$ are decreasing for increasing α, and the right endpoints, increasing ($[r]_\alpha$ denotes the equivalence class of r under the equivalence relation $\equiv_\alpha$). Since there is no strictly monotone sequence of real numbers of type ω_1, there is an ordinal $\beta_r < \omega_1$ such that both the left and right endpoints of $[r]_\alpha$ are constant for $\alpha > \beta_r$. Let $\gamma = \sup\{\beta_r : r \text{ rational}\}$. Then all of the equivalence classes are constant for $\alpha > \gamma$, contradiction. □

Corollary 1.7. *Let A be a subalgebra of the interval algebra on* $\mathbf{R}$. *Then A does not have an uncountable ideal independent subset.*

PROOF. Suppose that X is an uncountable ideal independent subset of A. Let $\langle a_\alpha : \alpha < \omega_1 \rangle$ be a one-one enumeration of some elements of X. For each $\alpha < \omega_1$ let $I_\alpha = \langle \{a_\beta : \beta < \alpha\} \rangle^{\rm Id}$. Clearly then $\langle I_\alpha : \alpha < \omega_1 \rangle$ is a strictly increasing sequence of ideals in B, contradicting 1.6. □

Finally, we are ready for the theorem. The main content of the theorem is from Todorčević [86], Theorem 12, as we mentioned.

Theorem 1.8. *There is a BA A of power 2^ω such that:*

(i) Ult$A$ *has, for each $\kappa < 2^\omega$, a discrete subspace of power κ, and A has an atomic homomorphic image B with κ atoms;*

(ii) Ult$A$ *has no discrete subspace of power 2^ω;*

(iii) If B is any homomorphic image of A, then there is a dense subset X of B such that there is a decomposition $X = Y \cup \bigcup_{i \in \omega} Z_i$ with Y the set of all atoms of B and for each $i \in \omega$, the set Z_i has the finite intersection property.

PROOF. Let E_0 and E_1 be as in Lemma 1.5. Without loss of generality, $0, 1 \notin E_0 \cup E_1$. For $i < 2$ let K_i be the linearly ordered set obtained from

[0,1] by replacing each element $r \in E_i$ by two new points $r^- < r^+$. Taking the order topology on K_i, we obtain a Boolean space, as is easily verified. In fact, K_i is homeomorphic to the Stone space of the interval algebra on $E_i \cup \{0\}$. Namely, the following function f from Ult(Intalg($E_i \cup \{0\}$)) into K_i is the desired homeomorphism. Take any $F \in$ Ult(Intalg($E_i \cup \{0\}$)). If there is an $r \in E_i$ such that $F \cap \{[0,a) : a \in E_i \cup \{1\}\} = \{[0,a) : r \le a\}$, we let $fF = r^-$. If there is an $r \in E_i$ such that $F \cap \{[0,a) : a \in E_i \cup \{1\}\} = \{[0,a) : r < a\}$, we let $fF = r^-$. In all other cases we let $fF = \inf\{a : a \in E_i \cup \{1\} : [0,a) \in F\}$. Clearly f is one-one and maps onto K_i. To show that it is continuous, first note that the following clopen subsets of K_i constitute a base for its topology:

$$\{[r^+, s^-) : r, s \in E_i,\ r < s\} \cup \{[0, s^-) : s \in E_i\} \cup \{[r^+, 1) : r \in E_i\}.$$

Then it is easy to check (with obvious assumptions) that

$$f^{-1}[[r^+, s^-)] = \{F : [r,s) \in F\};$$
$$f^{-1}[[0, s^-)] = \{F : [0,s) \in F\};$$
$$f^{-1}[[r^+, 1)] = \{F : [r,1) \in F\}.$$

This completes the proof that f is a homeomorphism from Ult(Intalg($E_i \cup \{0\}$)) into K_i.

By Corollary 1.7, neither K_0 nor K_1 has an uncountable discrete subspace. Also, $K_0 \times K_1$ is a Boolean space, and we let A be the BA of closed-open subsets of it.

First we check that for any $\kappa < 2^\omega$, $K_0 \times K_1$ has a discrete subset of power κ. Let f be a strictly increasing function from a subset of E_0 of power κ into E_1. Then we claim that $D \stackrel{\text{def}}{=} \{(r^-, (fr)^+) : r \in \text{dom} f\}$ is discrete. To show this, for each $r \in \text{dom} f$ let $a_r = [0, r^+) \times ((fr)^-, 1]$. Suppose $(s^-, (fs)^+) \in a_r$ and $s \neq r$. Thus $s^- < r^-$ and $(fr)^+ < (fs)^+$, contradiction.

From the proofs of Theorems 1.2 and 1.3 it now follows that A has a homomorphic image C which has a disjoint subset of power κ. By an easy application of the Sikorski extension theorem, A has an atomic homomorphic image B with κ atoms.

Next we prove (ii). Suppose that D is a discrete subspace of $K_0 \times K_1$ of size 2^ω. Now K_1 has no uncountable discrete subspace, so for each $x \in \text{dom} D$, the set $\{y : (x,y) \in D\}$ is countable. It follows that we may assume that D is a function. Similarly, we may assume that D is one-one.

For $(r,s) \in D$ let a_{rs} and b_{rs} be open intervals in K_0 and K_1 respectively such that $(a_{rs} \times b_{rs}) \cap D = \{(r,s)\}$. Let F_0 and F_1 be countable dense subsets of K_0 and K_1 respectively (in the sense that if $a < b$ in K_0 and $(a,b) \neq 0$ then there is a $c \in F_0$ such that $a < c < b$; similarly for K_1). Suppose that $\text{dom} D \backslash (\{r^- : r \in E_0\} \cup \{r^+ : r \in E_0\})$ has power 2^ω. Then we may successively assume that $\text{dom}D \cap (\{r^- : r \in E_0\} \cup \{r^+ : r \in E_0\})=0$, that each a_{rs} is an open interval with endpoints in F_0, and that all of the a_{rs} are equal, which implies that D has only one element, contradiction.

Thus we may assume that $\text{dom}D \subseteq \{r^- : r \in E_0\} \cup \{r^+ : r \in E_0\}$, and similarly for $\text{ran}D$. Hence we may assume that there are $\varepsilon, \delta \in \{0,1\}$ such that $\text{dom}D \subseteq \{r^\varepsilon : r \in E_0\}$ and $\text{ran}D \subseteq \{r^\delta : r \in E_1\}$.

Thus there are now four cases, which are very similar, and we treat only one of them: $\varepsilon = \delta = 0$. We may assume that for each $(r^-, s^-) \in D$ the right endpoint of $a_{r^-s^-}$ is r^+ and that of $b_{r^-s^-}$ is s^+. Furthermore, we may assume that there exist $q_i \in F_i$, $i = 0, 1$, such that $q_0 \in a_{r^-s^-}$ and $q_1 \in b_{r^-s^-}$ for each $(r^-, s^-) \in D$. Now we claim that the mapping $r \mapsto s$ for $(r^-, s^-) \in D$ is strictly decreasing (contradiction!). For, suppose that $(r^-, s^-) \in D$, $(u^-, v^-) \in D$, $r < u$, and $s < v$. Then it is clear that $(r^-, s^-) \in a_{u^-v^-} \times b_{u^-v^-}$, a contradiction (using the facts that $q_0 \in a_{r^-s^-} \cap a_{u^-v^-}$, and $q_1 \in b_{r^-s^-} \cap b_{u^-v^-}$).

Now we turn to the last part of the theorem. Suppose that B is a homomorphic image of A. Let F_i be a countable dense subset of E_i for $i = 0, 1$. Let $E_i^+ = \{r^+ : r \in E_i\}$, $E_i^- = \{r^- : r \in E_i\}$ for $i = 0, 1$. Now we are going to define some subsets $X^{\cdots}_{\cdots}$ of B indexed by various countable objects; each subset will satisfy the finite intersection property, and this will be obvious in each case. What is not so obvious is what these sets are good for. We show after defining them that their union with the set of atoms of B is dense in B, which is the desired conclusion of the theorem. It is convenient to work with the dual of B, which is some closed subspace Y of $K_0 \times K_1$.

Suppose that $p, q \in F_0$, $r, s \in F_1$, $p < q$, $r < s$, and $([p^+, q^-] \times [r^+, s^-]) \cap Y \neq 0$; then we set

$$X^1_{pqrs} = \{([p^+, q^-] \times [r^+, s^-]) \cap Y\}.$$

Next, suppose that $q \in F_0$, $r, s \in F_1$, and $r < s$. Then we set

$$X^2_{qrs} = \{([x, q^-] \times [r^+, s^-]) \cap Y : x \in E_0^+,\ x < q, \text{ and } \exists y(r^+ < y < s^- \text{ and } (x, y) \in Y)\}.$$

The next three sets are similar to X^2_{qrs}. Suppose that $p \in F_0$, $r, s \in F_1$, and $r < s$. Set

$$X^3_{prs} = \{([p^+, x] \times [r^+, s^-]) \cap Y : x \in E_0^-,\ p < x, \text{ and } \exists y(r^+ < y < s^- \text{ and } (x, y) \in Y)\}.$$

Suppose that $p, q \in F_0$, $s \in F_1$, and $p < q$. Set

$$X^4_{pqs} = \{([p^+, q^-] \times [y, s^-]) \cap Y : y \in E_1^+,\ y < s, \text{ and } \exists x(p^+ < x < q^- \text{ and } (x, y) \in Y)\}.$$

Suppose that $p, q \in F_0$, $r \in F_1$, and $p < q$. Set

$$X^5_{pqr} = \{([p^+, q^-] \times [r^+, y]) \cap Y : y \in E_1^-,\ r < y,\\ \text{and } \exists x(p^+ < x < q^- \text{ and } (x, y) \in Y)\}.$$

Now suppose that $p, q \in F_0$, $r, s \in F_1$, $p < q$, and $r < s$. Set

$$X^6_{pqrs} = \{([x, q^-] \times [y, s^-]) \cap Y : x \in E_0^+,\ y \in E_1^+,\ x < p^-,\\ y < r^-, \text{ and } ([p^+, q^-] \times [r^+, s^-]) \cap Y \neq 0\}.$$

The next three sets are similar to X^6_{pqrs}. For each of them we suppose that $p, q \in F_0$, $r, s \in F_1$, $p < q$, and $r < s$.

$$X^7_{pqrs} = \{([x, q^-] \times [r^+, y]) \cap Y : x \in E_0^+,\ y \in E_1^-,\ x < p^-,\\ s^+ < y, \text{ and } ([p^+, q^-] \times [r^+, s^-]) \cap Y \neq 0\}.$$

$$X^8_{pqrs} = \{([p^+, x] \times [y, s^-]) \cap Y : x \in E_0^-,\ y \in E_1^+,\ q^+ < x,\\ y < r^-, \text{ and } ([p^+, q^-] \times [r^+, s^-]) \cap Y \neq 0\}.$$

$$X^9_{pqrs} = \{([p^+, x] \times [r^+, y]) \cap Y : x \in E_0^-,\ y \in E_1^-,\ q^+ < x,\\ s^+ < y, \text{ and } ([p^+, q^-] \times [r^+, s^-]) \cap Y \neq 0\}.$$

Next, if $p \in F_0$ and $r, s \in F_1$ with $r < s$, we set

$$X^{10}_{prs} = \{([p^+, x] \times [r^+, y]) \cap Y : x \in E_0^-,\ y \in E_1^-,\ s^+ < y,\ p < x,\\ \text{and there is a } v \text{ such that } (x, v) \in Y \text{ and } r^+ < v < s^-\}.$$

The other sets are similar to this one; with obvious assumptions,

$$X^{11}_{prs} = \{([p^+, x] \times [y, s^-]) \cap Y : x \in E_0^-,\ y \in E_1^+,\ y < r^-,\ p < x,\\ \text{and there is a } v \text{ such that } (x, v) \in Y \text{ and } r^+ < v < s^-\}.$$

$$X^{12}_{qrs} = \{([x, q^-] \times [r^+, y]) \cap Y : x \in E_0^+,\ y \in E_1^-,\ s^+ < y,\ x < q,\\ \text{and there is a } v \text{ such that } (x, v) \in Y \text{ and } r^+ < v < s^-\}.$$

$$X^{13}_{qrs} = \{([x, q^-] \times [y, s^-]) \cap Y : x \in E_0^+,\ y \in E_1^+,\ y < r^-,\ x < q,\\ \text{and there is a } v \text{ such that } (x, v) \in Y \text{ and } r^+ < v < s^-\}.$$

$$X^{14}_{pqs} = \{([x, q^-] \times [y, s^-]) \cap Y : x \in E_0^+,\ y \in E_1^+,\ x < p^-,\ y < s,\\ \text{and there is a } u \text{ such that } (u, y) \in Y \text{ and } p^+ < u < q^-\}.$$

$$X^{15}_{pqr} = \{([x, q^-] \times [r^+, y]) \cap Y : x \in E_0^+,\ y \in E_1^-,\ x < r^-,\ r < y, \text{ and there is a } u \text{ such that } (u, y) \in Y \text{ and } p^+ < u < q^-\}.$$

$$X^{16}_{pqs} = \{([p^+, x] \times [y, s^-]) \cap Y : x \in E_0^-,\ y \in E_1^+,\ q^+ < x,\ y < s, \text{ and there is a } u \text{ such that } (u, y) \in Y \text{ and } p^+ < u < q^-\}.$$

$$X^{17}_{pqr} = \{([p^+, x] \times [r^+, y]) \cap Y : x \in E_0^-,\ y \in E_1^-,\ q^+ < x,\ r < y, \text{ and there is a } u \text{ such that } (u, y) \in Y \text{ and } p^+ < u < q^-\}.$$

Now we show that the union of these sets with the set of atoms of B is dense in B. Suppose that U is a non-zero element of B; we may assume that U has the form $((a, b) \times (c, d)) \cap Y$, and that it is not $\geq$ any atom of B. Fix an element (x, y) of U. We consider various possiblities.

Case 1. $x \notin E_0^- \cup E_0^+$ and $y \notin E_1^- \cup E_1^+$. Then clearly there exist p, q, r, s such that $(x, y) \in X^1_{pqrs} \subseteq U$.

Case 2. $x \in E_0^-$ and $y \notin E_1^- \cup E_1^+$. There are p, r, s such that $(x, y) \in X^3_{prs} \subseteq U$.

Case 3. $x \in E_0^+$ and $y \notin E_1^- \cup E_1^+$. There are q, r, s such that $(x, y) \in X^2_{qrs} \subseteq U$.

Case 4. $x \notin E_0^- \cup E_0^+$. Similar to above cases, using $X^1_{\ldots}$, $X^4_{\ldots}$, or $X^5_{\ldots}$.

Case 5. $x \in E_0^-$ and $y \in E_1^-$. Then it is easy to find $p \in F_0$, $r \in F_1$ so that $(x, y) \in [p^+, x] \times [r^+, y] \cap Y \subseteq U$. Now there are two subcases. *Subcase 5.1.* There is a $(u, v) \in Y$ such that $p^+ < u < x$ and $r^+ < v < y$. Then it is easy to find q, s so that $(x, y) \in X^9_{pqrs} \subseteq U$. *Subcase 5.2.* Otherwise, since we are assuming that U is not $\geq$ any atom of B, either there is a v such that $(x, v) \in Y$ and $r^+ < v < y$, or there is a u such that $(u, y) \in Y$ and $p^+ < u < x$. In the first instance there is an s such that $(x, y) \in X^{10}_{prs} \subseteq U$. In the second instance we use X^{17}.

Case 6. $x \in E_0^-$ and $y \in E_1^+$. This is like Case 5. We use X^8, X^{16}, and X^{11}.

Case 7. $x \in E_0^+$ and $y \in E_1^-$. This is like Case 5. We use X^7, X^{15}, and X^{12}.

Case 8. $x \in E_0^+$ and $y \in E_1^+$. This is like Case 5. We use X^6, X^{14}, and X^{13}. □

Corollary 1.9. *Assume that 2^ω is a limit cardinal. Then there is a BA A of power 2^ω with spread 2^ω not attained, such that A has no homomorphic image B such that cB=sA.*

PROOF. The first part of the conclusion follows immediately from the theorem. Now suppose that B is a homomorphic image of A such that cB = sA. Since sA is not attained, it follows from Corollary 1.4 that cB is not attained. Now let X, Y, etc., be as in (iii) of the theorem. Then $|Y| < 2^\omega$ since cB is not attained. Let W be a disjoint subset of B of power $|Y|^+$. For each $w \in W$ choose $x_w \in X$ such that $x_w \leq w$, and let $X' = \{x_w : w \in W\}$. Then there has to exist an $i \in \omega$ such that $|X' \cap Z_i| > 2$, which is a contradiction, since X' is disjoint and Z_i has the finite intersection property. □

Returning to the program described in the introduction, we note that it is obvious that $\mathrm{cs_s}A = [\omega, \mathrm{c}A]$. The caliber notion associated with cellularity has been worked on a lot. There are several variants of this notion. For a survey of results and problems, see Comfort, Negrepontis [82]. We shall compare c with other cardinal functions one-by-one in the discussion of those functions.

We turn to the relation $\mathrm{cs_r}$; see the end of the introduction. We do not have a purely cardinal number characterization of this relation. Some restrictions to put on it are given in the following simple theorem:

Theorem 1.10. *For any infinite BA A the following conditions hold:*
(i) If $(\kappa, \lambda) \in \mathrm{cs_r}A$, *then* $\kappa \leq \lambda \leq |A|$ *and* $\kappa \leq \mathrm{c}A$.
(ii) For each $\kappa \in [\omega, \mathrm{c}A]$ *we have* $(\kappa, \kappa) \in \mathrm{cs_r}A$.
(iii) If $(\kappa, \lambda) \in \mathrm{cs_r}A$ *and* $\kappa \leq \mu \leq \lambda$, *then* $(\kappa, \mu) \in \mathrm{cs_r}A$.
(iv) If $(\lambda, (2^\kappa)^+) \in \mathrm{cs_r}A$ *for some* $\lambda \leq \kappa$, *then* $(\omega, (2^\kappa)^+) \in \mathrm{cs_r}A$.
(v) $(\mathrm{c}A, |A|) \in \mathrm{cs_r}A$. □

The proof of this theorem is easy; for (iv), use Theorem 10.1 of Part I of the Handbook. To understand more about the possiblities for the relation $\mathrm{cs_r}A$, consider the following examples. If κ is an infinite cardinal and A is the finite-cofinite algebra on κ, then $\mathrm{cs_r}A = \{(\lambda, \lambda) : \lambda \in [\omega, \kappa]\}$. If A is the free algebra on κ free generators, then $\mathrm{cs_r}A = \{(\omega, \lambda) : \lambda \in [\omega, \kappa]\}$. If A is an infinite interval algebra and we assume GCH, then $\mathrm{cs_r}A$ does not have any gaps of size 2 or greater. That is, if $(\kappa, \lambda) \in \mathrm{cs_r}A$, then $\lambda = \kappa$ or $\lambda = \kappa^+$. This is seen by using Theorem 10.1 again: such a gap would imply the existence in A of an uncountable independent subset, which does not exist in an interval algebra. There are two deeper results:

(1) Todorčević in [87] shows that it is consistent to have for each regular non-weakly compact cardinal κ a κ-cc BA A of size κ such that any subalgebra or homomorphic image B of A of size $< \kappa$ has a disjoint family of size $|B|$. Applying this to subalgebras and to non-limit cardinals, this means in our terminology that is is consistent to have an algebra A with $\mathrm{cs_r}A = \{(\lambda, \lambda) : \lambda \in [\omega, \kappa]\} \cup \{(\kappa, \kappa^+)\}$.

(2) In models of Kunen [78] and Foreman, Laver [86], every ω_2-cc algebra of size ω_2 contains an ω_1-cc subalgebra of size ω_1. Thus in these models certain relations $\mathrm{cs_r}$ are ruled out; cf. (1).

There are many problems remaining concerning the relations $\mathrm{cs_r}$. Considering small cardinals—$\omega, \omega_1, \omega_2$—one can start cataloging known relations and open problems. We mention one such problem.

Problem 2. *Is there a BA A with* $\mathrm{cs_{Sr}}A = \{(\omega,\omega), (\omega,\omega_1), (\omega_1,\omega_1), (\omega,\omega_2), (\omega_2,\omega_2)\}$*? Equivalently, is there a BA A such that* $|A| = \omega_2 = \mathrm{c}A$*, A has a ccc subalgebra of power* ω_2*, and every subalgebra of A of size* ω_2 *either has cellularity* ω *or* ω_2*?*

The relation $\mathrm{c_{Hr}}A$ is similar to $\mathrm{cs_{Sr}}A$:

Theorem 1.11. *For any infinite BA A the following conditions hold:*
(i) If $(\kappa,\lambda) \in \mathrm{c_{Hr}}A$*, then* $\kappa \le \lambda \le |A|$ *and* $\kappa \le \mathrm{s}A$*.*
(ii) For each $\kappa \in [\omega, \mathrm{s}A)$ *there is a* $\lambda \le 2^\kappa$ *such that* $(\kappa,\lambda) \in \mathrm{c_{Hr}}A$*.*
(iii) If $(\lambda,(2^\kappa)^+) \in \mathrm{c_{Hr}}A$*, for some* $\lambda \le \kappa$*, then* $(\omega,(2^\kappa)^+) \in \mathrm{c_{Hr}}A$*.*
(iv) $(\mathrm{c}A, |A|) \in \mathrm{c_{Hr}}A$*.*

PROOF. Only (ii) needs some comment. Let $\kappa \in [\omega, \mathrm{s}A)$. Take a homomorphic image B of A such that $\mathrm{c}B > \kappa$; let C be a subalgebra of B generated by a disjoint set of power κ, and extend the identity on C to a homomorphism from B onto a subalgebra D of $\overline{C}$; then D is as desired. □

Note in Theorem 1.11 (ii) that $\kappa = \mathrm{s}A$ is not in general possible, by Corollary 1.9. The following examples shed some light on $\mathrm{c_{Hr}}$. If A is complete and $(\kappa,\lambda) \in \mathrm{c_{Hr}}A$, then $\lambda^\omega = \lambda$. If A is the finite-cofinite algebra on an infinite cardinal κ, then $\mathrm{c_{Hr}}A = \{(\lambda,\lambda) : \omega \le \lambda \le \kappa\}$. If A is the free BA on κ free generators, κ infinite, then $\mathrm{c_{Hr}}A = \{(\lambda,\mu) : \omega \le \lambda \le \mu \le \kappa\}$. If A an infinite interval algebra, then there is no gap of size 2 or greater in $\mathrm{c_{Hr}}A$, in the same sense as above. The algebra A of Todorčević [87] has $\mathrm{c_{Hr}}A = \{(\lambda,\lambda) : \lambda \in [\omega,\kappa]\} \cup \{(\kappa,\kappa^+)\}$. Another example is $\mathcal{P}\omega$. Under CH, its homomorphic cellularity relation is $(\omega,\omega_1), (\omega_1,\omega_1)$. If we assume that $2^\omega = \omega_2$ then we see that its homomorphic cellularity relation is $\{(\omega,\omega_2), (\omega_1,\omega_2), (\omega_2,\omega_2)\}$. Another relevant result is from Koppelberg [77]: assuming MA, if A is an infinite BA with $|A| < 2^\omega$, then A has a countable homomorphic image. And a special case of a result of Just, Koszmider [87] is that it is consistent to have $2^\omega = \omega_2$ with an algebra A having homomorphic cellularity relation $\{(\omega,\omega_1), (\omega_1,\omega_1)\}$. Finally, Fedorchuk [75] constructed, assuming $V = L$, a BA A such that $\mathrm{c_{Hr}}A = \{\omega,\omega_1\}$.

Again, if one starts to systematically study c_{Hr} for small cardinals, one comes to many open problems. We mention one of them.

Problem 3. Is there a BA with c_{Hr} relation $\{(\omega,\omega), (\omega,\omega_1), (\omega_1,\omega_2)\}$?

To conclude this section, we consider cellularity for special classes of BA's. For an atomic BA A, $\mathrm{c}A$ coincides with the number of atoms of A. Also note that some of the free product questions are trivial for atomic algebras; in particular, $\mathrm{c}(A \oplus B) = \max\{\mathrm{c}A, \mathrm{c}B\}$ if A and B are atomic. There is one interesting result which comes up in considering cellularity and unions for complete BA's; this result is evidently due to Solovay, Tennenbaum [71]:

Theorem 1.12. *Let κ and λ be uncountable regular cardinals, and suppose that $\langle A_\alpha : \alpha < \lambda\rangle$ is an increasing sequence of complete BA's satisfying the κ-chain condition, such that A_α is a complete subalgebra of A_β for $\alpha < \beta < \lambda$, and for γ infinite $< \lambda$, $\bigcup_{\alpha<\gamma} A_\alpha$ is dense in A_γ. Then $\bigcup_{\alpha<\lambda} A_\alpha$ also satisfies the κ-chain condition.*

PROOF. By the discussion above of chain conditions in unions, we may assume that $\kappa = \lambda$. Let $B = \bigcup_{\alpha<\kappa} A_\alpha$. For each $\alpha < \kappa$ we define c_α mapping B into A_α by setting

$$c_\alpha x = \prod_{x \leq a \in A_\alpha} a.$$

(This function is a *cylindrification* on B, but we do not need to check that.)

Now, in order to get a contradiction, assume that X is a disjoint subset of B of size $\geq \kappa$. We may assume that X is maximal disjoint. Take any $\alpha < \kappa$. Now $\sum X = 1$, and hence $\sum\{c_\alpha x : x \in X\} = 1$. Since each A_α satisfies the κ-chain condition, choose $X_\alpha \subseteq X$ of size $< \kappa$ such that

(1) $\sum\{c_\alpha x : x \in X_\alpha\} = 1$.

Choose $\beta_\alpha < \kappa$ such that $X_\alpha \subseteq A_{\beta_\alpha}$; the ordinal β_α exists since $|X_\alpha| < \kappa$ and κ is regular. Finally, let γ be a limit ordinal $< \kappa$ such that $\beta_\alpha < \gamma$ for all $\alpha < \gamma$; the existence of γ is easy to see. We shall now prove that $X \subseteq A_\gamma$ (contradiction!).

Let $x \in X$ be arbitrary. Since $\bigcup_{\alpha<\gamma} A_\alpha$ is dense in A_γ, choose a non-zero $b \in \bigcup_{\alpha<\gamma} A_\alpha$ such that $b \leq c_\gamma x$. Say $b \in A_\alpha$ with $\alpha < \gamma$. By (1), choose $a \in X_\alpha$ such that $c_\alpha a \cdot b \neq 0$. If $b \cdot a = 0$, then $a \leq -b$ and hence $c_\alpha a \leq -b$ and so $c_\alpha a \cdot b = 0$, contradiction. Thus $b \cdot a \neq 0$, and so $c_\gamma x \cdot a \neq 0$. It follows that $x \cdot a \neq 0$, by the same argument as above. But both x and a are in X, so $x = a$. Thus $x \in A_\alpha \subseteq A_\gamma$, as desired. □

There is a large literature on cellularity for BA's of the form $\mathcal{P}(\kappa)/I$; for a start, see Baumgartner, J., Taylor, A., Wagon, S. [82]. Usually BA terminology is not used in such investigations; *saturation* of ideals is the term used.

2. DEPTH

Recall that Depth$A$ is the supremum of cardinalities of subsets of A which are well-ordered by the Boolean ordering. There are two main references for results about this notion: McKenzie, Monk [82] and (implicitly) Grätzer, Lakser [69]. (The paper of Grätzer and Lakser was not known to us until recently. Theorems 1.2.4 and 1.3.2 and their corollaries, all in their versions for length, had already appeared in the paper by Grätzer and Lakser.)

Some of the results which we shall present about depth depend on the following simple lemma.

Lemma 2.1. *Let A and B be BA's, and let X be a chain in $A \times B$ of infinite cardinality κ. Then the projections of X are chains, and at least one of them has cardinality κ. Furthermore, if X has order type κ, then X has a subset of order type κ on which one of the two projections is one-one.*

PROOF. For any $z \in X$ write $z = (z_0, z_1)$. For $i = 0, 1$ write $z \equiv_i w$ iff $z, w \in X$ and $z_i = w_i$. Now note that

$$\{\{x\} : x \in X\} \subseteq \bigcup\{a \cap b : a \in X/\equiv_0,\ b \in X/\equiv_1\};$$

hence one of the two equivalence relations $\equiv_0, \equiv_1$ has κ equivalence classes, and the lemma follows. □

Now we shall show that Depth$A$ is attained if Depth$A$ is a successor cardinal or a cardinal of cofinality ω; otherwise, there are counterexamples.

Theorem 2.2. *If* cf(Depth$A$) $= \omega$, *then* Depth$A$ *is attained.*

PROOF. Let $\kappa =$ Depth$A$. We may assume that $\kappa > \omega$. Let $\langle \lambda_i : i < \omega \rangle$ be a strictly increasing sequence of cardinals with supremum κ, and with $\lambda_0 = 0$. Now we call an element a of A an $\infty-$element if λ_i is embeddable in $A \upharpoonright a$ for all $i < \omega$. We claim

(*) If a is an ∞-element, and $a = b + c$ with $b \cdot c = 0$, then b is an ∞-element or c is an ∞-element.

In fact, by Lemma 2.1, for each $i < \omega$, λ_i is embeddable in $A \upharpoonright b$ or $A \upharpoonright c$, so (*) follows.

Using (*), we construct a sequence $\langle a_i : i < \omega \rangle$ of elements of A by induction. Suppose that a_j has been constructed for all $j < i$ so that $b \stackrel{\text{def}}{=} \prod_{j<i} -a_j$ is an ∞-element. Let $\langle c(\alpha) : \alpha < \lambda_{i+1} \rangle$ be an isomorphic embedding of λ_{i+1} into b. By (*), one of the elements $c(\lambda_i)$ and $b \cdot -c(\lambda_i)$ is an ∞-element, while clearly λ_i is embeddable in both of these elements. So we can choose

$a_i \leq b$ so that λ_i is embeddable in a_i, and $\prod_{j\leq i} -a_j$ is an ∞-element. This finishes the construction.

For each $i < \omega$ let $\langle b_{i\alpha} : \alpha < \lambda_i \rangle$ be an embedding of λ_i into a_i. Note that $a_i \cdot a_j = 0$ for $i < j < \omega$. Hence the following sequence $\langle d_\alpha : \alpha < \kappa \rangle$ is clearly the desired embedding of κ into A. Given $\alpha < \kappa$, there is a unique $i < \omega$ such that $\lambda_i \leq \alpha < \lambda_{i+1}$. We let $d_\alpha = a_0 + \cdots + a_i + b_{i+1,\alpha}$. □

In order to see that Theorem 2.2 is "best possible", it is convenient to first discuss the depth of products.

Theorem 2.3. $\text{Depth}\prod_{i\in I} A_i = \max(|I|, \sup_{i\in I}\text{Depth}A_i)$.

PROOF. Clearly $\geq$ holds. Suppose $=$ fails to hold, and let f be an order isomorphism of κ^+ into $\prod_{i\in I} A_i$, where $\kappa = \max(|I|, \sup_{i\in I}\text{Depth}A_i)$. For each $i \in I$ there is an ordinal $\alpha_i < \kappa^+$ such that $(f\alpha_i)_i = (f\beta)_i$ for all $\beta > \alpha_i$. Let $\gamma = \sup_{i\in I}\alpha_i$. Then for all $\delta > \gamma$ we have $f\delta = f\gamma$, contradiction. □

Theorem 2.4. *Let $\kappa = \sup_{i\in I}\text{Depth}A_i$, and suppose that κ is regular. Then the following conditions are equivalent:*

(i) $\text{Depth}\prod_{i\in I} A_i$ *is not attained.*

(ii) $|I| < \kappa$, *and for all $i \in I$, A_i has no chain of order type κ.* □

The proof of this theorem is very similar to that of Theorem 2.3. The case of singular cardinals is a little more involved:

Theorem 2.5. *Let $\kappa = \sup_{i\in I}\text{Depth}A_i$, and suppose that κ is singular. Then the following conditions are equivalent:*

(i) $\text{Depth}\prod_{i\in I} A_i$ *is not attained.*

(ii) These four conditions hold:

(a) $|I| < \kappa$.

(b) For all $i \in I$, A_i has no chain of type κ.

(c) $|\{i \in I : \text{Depth}A_i = \kappa\}| < \text{cf}\kappa$.

(d) $\sup\{\text{Depth}A_i : i \in I, \text{Depth}A_i < \kappa\} < \kappa$.

PROOF. (i)$\Rightarrow$(ii) is clear. Assume (ii), and suppose that $\langle x_\alpha : \alpha < \kappa \rangle$ is strictly increasing in $\prod_{i\in I} A_i$. Define

$$J_i = \{\alpha < \kappa : x_\alpha i < x_{\alpha+1} i\} \text{ for } i \in I;$$
$$K = \{i \in I : \text{Depth}A_i = \kappa\};$$
$$\lambda = \sup\{\text{Depth}A_i : i \in I, \text{Depth}A_i < \kappa\}.$$

Then by the above assumptions we have $\lambda < \kappa$, $|J_i| \leq \lambda$ for all $i \in I\backslash K$, $|K| < \text{cf}\kappa$, and $|J_i| < \kappa$ for all $i \in K$. It follows that $|\bigcup_{i\in I} J_i| < \kappa$. But for any $\alpha \in \kappa\backslash\bigcup_{i\in I} J_i$ we have $x_\alpha = x_{\alpha+1}$, contradiction. □

The above theorems completely describe the depth of products. The case of weak products is even simpler:

Theorem 2.6. *Let* $\kappa = \sup_{i\in I}\mathrm{Depth}A_i$, *and suppose that* $\mathrm{cf}\kappa > \omega$. *Then the following conditions are equivalent:*
(i) $\prod^{\mathrm{w}}_{i\in I} A_i$ *has no chain of order type* κ.
(ii) For all $i \in I$, A_i *has no chain of order type* κ.

PROOF. Obviously $(i) \Rightarrow (ii)$. $(ii) \Rightarrow (i)$: Suppose that $\langle x_\alpha : \alpha < \kappa\rangle$ is strictly increasing in $\prod^{\mathrm{w}}_{i\in I} A_i$. For any $y \in \prod^{\mathrm{w}}_{i\in I} A_i$ let $Sy = \{i \in I : y_i \neq 0\}$.
Case 1. Sx_α is finite for all $\alpha < \kappa$. Since $\mathrm{cf}\kappa > \omega$, it follows that there is an $\alpha < \kappa$ such that $Sx_\alpha = Sx_\beta$ whenever $\alpha < \beta < \kappa$. But then Lemma 2.1 easily gives a contradiction.
Case 2. Otherwise we may assume that $\{i \in I : x_\alpha i \neq 1\}$ is finite for all $\alpha < \kappa$, and a contradiction is reached as in Case 1. □

Corollary 2.7. $\mathrm{Depth}\prod^{\mathrm{w}}_{i\in I} A_i = \sup_{i\in I}\mathrm{Depth}A_i$. □

Theorem 2.6 enables us to easily show that Theorem 2.2 is best possible: if κ is a limit cardinal with $\mathrm{cf}\kappa > \omega$, then it is easy to construct a weak product B such that $\mathrm{Depth}B = \kappa$ but depth is not attained in B.

If A is a subalgebra of B, then obviously $\mathrm{Depth}A \leq \mathrm{Depth}B$ and the difference can be arbitrarily large. If A is a homomorphic image of then depth can change either way from A to B; see the argument here for cellularity.

For ultrafilters, the situation is also similar to that for cellularity. The same argument as before shows that if F is a countably complete ultrafilter on an infinite set I and is a BA with depth ω for each $i \in I$, then $\prod_{i\in I} A_i/F$ has depth ω. And, as before, if F is a countably incomplete ultrafilter on I and each algebra is infinite, then $\prod_{i\in I} A_i/F$ has depth $> \omega$. This is easiest to see by recalling that $\prod_{i\in I} A_i/F$ is ω_1-saturated, and noting

(*) *If an infinite BA A is* κ-*saturated, then A has a chain of order type* κ.

To prove (*), we construct $a \in {}^\kappa A$ by recursion. Suppose that a_β has been defined for all $\beta < \alpha$, so that if β is a successor ordinal $\gamma + 1$, then $A \restriction -a_\gamma$ is infinite. If β is a successor ordinal, it is clear how to proceed in order to still have the indicated condition. If β is limit, consider the set

$$\{c_{x_\alpha} < v_0 : \alpha < \beta\} \cup \{\text{ “there are at least } n\text{”}v_1(v_0 < v_1) : n \in \omega\}.$$

This set is finitely satisfiable in A, and so an element satisfying all of these formulas gives the desired element a_β.

The third result which we gave for cellularity of ultraproducts, involving the notion of regularity of ultraproducts, does not extend so easily; in fact, we do not know whether it holds for depth. We give two results relevant to this. The first result follows easily from a theorem of W. Hodges, that if F is a regular ultrafilter on I then in ${}^I\langle\omega, >\rangle/F$ there is a chain of order type $|I|^+$. We give a direct BA proof of the BA result:

Theorem 2.8. *Let F be a $|I|$-regular ultrafilter on I, and suppose that A_i is an infinite BA for every $i \in I$. Then in $\prod_{i\in I} A_i/F$ there is a chain of order type $|I|^+$.*

PROOF. For brevity set $\kappa = |I|$. By the definition of regularity choose $E \subseteq F$ such that $|E| = \kappa$ and for all $i \in I$ the set $\{e \in E : i \in e\}$ is finite. Let G be a one-one function from E onto κ. For each $i \in I$ choose a strictly increasing sequence $\langle x_{ij} : j < \omega\rangle$ in A_i, and let $X_i = \{x_{ij} : j < \omega\}$. Then it suffices to show:

(*) If $g_\alpha \in \prod_{i\in I} X_i$ for all $\alpha < \kappa$, then there is an $f \in \prod_{i\in I} X_i$ such that $g_\alpha/F < f/F < 1$ for all $\alpha < \kappa$.

To define f, let $i \in I$. Let $e(1), \ldots, e(m)$ be all of the elements u of E such that $i \in u$. Then let fi be any element of X_i greater than all of the elements $g_{Ge(1)}i, \ldots, g_{Ge(m)}i$. This defines f. Now if $\alpha < \kappa$ and $i \in G^{-1}\alpha$, we have $g_\alpha i < fi < 1$, as desired. □

Theorem 2.9. *Let I be an infinite set, and suppose that A_i is an infinite BA for every $i \in I$. Then there is a proper filter G on I such that G contains all cofinite sets, and for every ultrafilter F including G, $\prod_{i\in I} A_i/F$ has a chain of order type $2^{|I|}$.*

PROOF. Again let $\kappa = |I|$. Let $S \subseteq {}^\kappa\omega$ satisfy the following condition:

(1) $|S| = 2^\kappa$, and for every finite sequence $i_0, \ldots, i_{k-1}$ of natural numbers and every sequence $f_0, \ldots, f_{k-1}$ of distinct members of S of length k, there is an $\alpha < \kappa$ such that $f_t\alpha = i_t$ for all $t < k$.

For the existence of such a set, see Comfort, Negrepontis [74], pp. 75-77. Let $\langle f_\alpha : \alpha < 2^\kappa\rangle$ enumerate S without repetitions. For $\alpha < \beta < 2^\kappa$, let $J_{\alpha\beta} = \{\gamma < \kappa : f_\alpha\gamma < f_\beta\gamma\}$. From (1) it is clear that the intersection of any finite number of the sets $J_{\alpha\beta}$ is infinite. Hence

$$\{J_{\alpha\beta} : \alpha < \beta < 2^\kappa\} \cup \{\Gamma \subseteq \kappa : |\kappa\backslash\Gamma| < \omega\}$$

generates a proper filter G containing all cofinite sets. Clearly G is as desired. □

These results leave the following problem open:

Problem 4. *Is the following conjecture true? If I is an infinite set, F is a $|I|$-regular ultrafilter on I, and A_i is an infinite BA for every $i \in I$, then in $\prod_{i\in I} A_i/F$ there is a chain of order type $2^{|I|}$.*

For free products, we have $\text{Depth}(\oplus_{i\in I}A_i) = \sup_{i\in I}\text{Depth}A_i$. The proof is somewhat involved, and will be omitted; see McKenzie, Monk [82].

Concerning unions, we note that Depth is an ordinary sup function with respect to the function P, where $PA = \{X \subseteq A : X \text{ is a well-ordered chain in } A\}$, and so Theorem 1.1 applies.

Next we discuss derived functions with respect to depth. The first result is that $\text{Depth}_{\text{H}+}$ is the same as tightness. To prove this, we need an equivalent form of tightness due to Arhangelskiĭ. It involves the notion of a free sequence in a topological space. Let X be a topological space. A *free sequence* in X is a sequence $\langle x_\xi : \xi < \alpha\rangle$ (α an ordinal) of elements of X such that for all $\xi < \alpha$ we have $\overline{\{x_\eta : \eta < \xi\}} \cap \overline{\{x_\eta : \xi \leq \eta < \alpha\}} = 0$. For an arbitrary topological space X and a point $x \in X$, the *tightness* $\text{t}x$ of x in X is, by definition, the least cardinal κ such that if $Y \subseteq X$ and $x \in \overline{Y}$, then there is a subset $Z \subseteq Y$ such that $|Z| \leq \kappa$ and $x \in \overline{Z}$. And the tightness $\text{t}X$ of X itself is $\sup_{x\in X}\text{t}x$. Clearly this means that $\text{t}A = \text{t}(\text{Ult}A)$ for any BA A. Arhangelskiĭ's equivalent form of tightness is given in the following theorem.

Theorem 2.10. *Let X be a compact Hausdorff space. Then* $\text{t}X = \sup\{|\alpha| :$ *there is a free sequence in X of order type* $\alpha\}$.

PROOF. Let $\text{t}X = \kappa$. In order to prove $\geq$, suppose to the contrary that $\langle x_\xi : \xi < \kappa^+\rangle$ is a free sequence. Then

(1) There is a $y \in X$ such that $|U \cap \{x_\xi : \xi < \kappa^+\}| = \kappa^+$ for every neighborhood U of y.

In fact, otherwise for every $y \in X$ let $U(y)$ be an open neighborhood of y such that $|U(y) \cap \{x_\xi : \xi < \kappa^+\}| \leq \kappa$. Thus $\{U(y) : y \in X\}$ is an open cover of X. Let $U(y_0), \ldots, U(y_{n-1})$ be a finite subcover. Then

$$\{x_\xi : \xi < \kappa^+\} = \bigcup_{i<n}(U(y_i) \cap \{x_\xi : \xi < \kappa^+\}),$$

and the right side has cardinality $\leq \kappa$, contradiction. So (1) holds.

Take y as in (1). Thus $y \in \overline{\{x_\xi : \xi < \kappa^+\}}$. Hence by the definition of tightness, choose a subset Γ of κ^+ of power at most κ such that $y \in \overline{\{x_\xi : \xi \in \Gamma\}}$. Let $\eta = \sup\Gamma + 1$. Hence $y \in \overline{\{x_\xi : \xi < \eta\}}$, so by freeness $y \notin \overline{\{x_\xi : \eta \leq \xi\}}$. So there is a neighborhood U of y such that $U \cap \{x_\xi : \eta \leq \xi\} = 0$. This contradicts (1). We have now proved $\geq$ in the theorem.

Now, for $\leq$, suppose that $1 \leq \lambda < \kappa$. We shall construct a free sequence of length λ^+. Choose $y \in X$ with $\text{t}(y) > \lambda$; say $Y \subseteq X$, $y \in \overline{Y}$, and for all $Z \subseteq Y$ with $|Z| \leq \lambda$, $y \notin \overline{Z}$. Set

$$Y' = \{x : \text{there is a } Z \subseteq Y \text{such that } |Z| \leq \lambda \text{ and } x \in \overline{Z}\}.$$

Thus $Y \subseteq Y'$, so $y \in \overline{Y'}$. Note

(2) If $Z \subseteq Y'$ and $|Z| \leq \lambda$, then $y \notin \overline{Z}$;
(3) If $Z \subseteq Y'$, $|Z| \leq \lambda$, and $z \in \overline{Z}$, then $z \in Y'$.

We now construct x_ξ, F_ξ, U_ξ for $\xi < \lambda^+$ such that $x_\xi \in Y'$, $y \in F_\xi \subseteq U_\xi$ with U_ξ open and F_ξ a closed neighborhood of y, by recursion. Suppose these have been constructed for all $\eta < \xi$, where $\xi < \lambda^+$. Since $y \notin \overline{\{x_\eta : \eta < \xi\}}$, let U_ξ be an open neighborhood of y such that $U_\xi \cap \{x_\eta : \eta < \xi\} = 0$. Let F_ξ be a closed neighborhood of y such that $F_\xi \subseteq U_\xi$. Then we claim

(4) $Y' \not\subseteq \bigcup_{\eta \leq \xi}(X \backslash F_\eta) \cup \overline{\{x_\eta : \eta < \xi\}}$.

For, suppose not; then we show that $y \in \overline{\{x_\eta : \eta < \xi\}}$ (contradiction). For, let U be an open neighborhood of y and let F' be a closed neighborhood of y which is included in U. Let W be the closure of the set $\{F_\eta : \eta \leq \xi\} \cup \{F'\}$ under finite intersections. Since $y \in \overline{Y'}$, for all $H \in W$ choose $z_H \in Y' \cap H$. Then $H' \cap \{z_H : H \in W\} \neq 0$ for all $H' \in W$. Choose

$$t \in \bigcap_{H \in W} H \cap \overline{\{z_H : H \in W\}}.$$

By (3), $t \in Y'$. Now $t \in F_\eta$ for all $\eta \leq \xi$, so by the "suppose not" for (4), $t \in \overline{\{x_\eta : \eta < \xi\}}$. Since $t \in F' \subseteq U$, it follows that $U \cap \{x_\eta : \eta < \xi\} \neq 0$, as desired.

So (4) holds; choose x_ξ in the left side of (4) but not in the right side. This completes the construction.

Suppose $\xi < \lambda^+$ and $s \in \overline{\{x_\eta : \eta < \xi\}} \cap \overline{\{x_\eta : \xi \leq \eta < \lambda^+\}}$. Then $s \notin U_\xi$, so $s \notin F_\xi$. Thus $s \in X \backslash F_\xi$, which is open, so there is an η with $\xi \leq \eta < \lambda^+$ such that $x_\eta \in X \backslash F_\xi$, contradiction. □

Theorem 2.11. *For any infinite BA A we have* $\mathrm{Depth}_{\mathrm{H+}} A = \mathrm{t}A$.

PROOF. For $\geq$, let $\langle F_\xi : \xi < \alpha \rangle$ be a free sequence; we produce a quotient A/I of A having a strictly increasing sequence of order type α. For brevity let $Y = \{F_\xi : \xi < \alpha\}$. For every $\xi < \alpha$ there is an element a_ξ of A such that $\{F_\eta : \eta < \xi\} \subseteq \mathcal{S}a_\xi$ and $\mathcal{S}a_\xi \cap \{F_\eta : \xi \leq \eta < \alpha\} = 0$. Consider the following ideal on A: $I = \{x \in A : Y \subseteq \mathcal{S}(-x)\}$. Suppose $\xi < \eta < \alpha$. Then $\mathcal{S}(a_\xi \cdot -a_\eta) \cap Y = 0$: if $F_\nu \in \mathcal{S}(a_\xi \cdot -a_\eta)$, then $F_\nu \in \mathcal{S}a_\xi$, hence $\nu < \xi$, and $-a_\eta \in F_\nu$, hence $\eta \leq \nu$, so $\eta < \xi$, contradiction. This shows that $[a_\xi] \leq [a_\eta]$ for $\xi < \eta < \alpha$. Still suppose that $\xi < \eta < \alpha$. Then $F_\xi \in \mathcal{S}a_\eta \backslash \mathcal{S}a_\xi = \mathcal{S}(a_\eta \cdot -a_\xi)$. Thus $Y \not\subseteq \mathcal{S}(-a_\eta + a_\xi)$, so $a_\eta \cdot -a_\xi \notin I$, which means that $[a_\eta] < [a_\xi]$, as desired.

For $\leq$, let I be an ideal in A, and let $\langle [a_\xi] : \xi < \alpha \rangle$ be a strictly increasing sequence in A/I. For each $\xi < \alpha$, the set $\{x : -x \in I\} \cup \{a_{\xi+1}, -a_\xi\}$ has the finite intersection property, since $a_{\xi+1} \cdot -a_\xi \notin I$. Let F_ξ be an ultrafilter including this set. Then, we claim, $\langle F_\xi : \xi < \alpha \rangle$ is a free sequence. To prove this it suffices to show that for any $\xi < \alpha$ we have

(1) $\{F_\eta : \eta < \xi\} \subseteq \mathcal{S}a_\xi$ and $\mathcal{S}a_\xi \cap \{F_\eta : \xi \leq \eta < \alpha\} = 0$.

If $\eta < \xi < \alpha$, then $a_{\eta+1} \cdot -a_\xi \in I$, and hence $-a_{\eta+1} + a_\xi \in F_\eta$; but also $a_{\eta+1} \in F_\eta$, so $a_\xi \in F_\eta$ and so $F_\eta \in \mathcal{S}a_\xi$, proving the first part of (1). For the second part, suppose that $\xi \leq \eta < \alpha$ and $F_\eta \in \mathcal{S}a_\xi$. Now $a_\xi \cdot -a_\eta \in I$, so $-a_\xi + a_\eta \in F_\eta$; but also $-a_\eta \in F_\eta$, so $-a_\xi \in F_\eta$, contradiction. □

Corollary 2.12. $\mathrm{Depth}_{\mathrm{H}+}$ *and* t *(for free sequences) have the same attainment properties, i.e., for any BA A and any infinite cardinal* κ, *A has a homomorphic image with a chain of order type* κ *iff* Ult*A has a free sequence of type* κ. □

Note that, as in the relation between spread and cellularity, $\mathrm{Depth}_{\mathrm{H}+}$ involves two sups, while t for free sequences involves only one; we return to this below.

Since Depth$A \leq$ cA, it is clear that $\mathrm{Depth}_{\mathrm{H}-}A = \omega$. It is also easy to see that $\mathrm{Depth}_{\mathrm{S}+}A =$ Depth$A$ and $\mathrm{Depth}_{\mathrm{S}-}A = \omega$. $\mathrm{Depth}_{\mathrm{h}+}$ is a little more interesting:

Theorem 2.13. $\mathrm{Depth}_{\mathrm{h}+}A =sA$ *for any infinite BA A.*

PROOF. For $\geq$, suppose that Y is a discrete subspace of Ult$A$; clearly Y, since it is discrete, has an increasing sequence of closed-open sets of order type $|Y|$. For $\leq$, suppose that Y is a subspace of Ult$A$ and $\langle U_\alpha : \alpha < \kappa \rangle$ is a strictly increasing system of closed-open subsets of Y. For each $\alpha < \kappa$ choose $y_\alpha \in U_{\alpha+1} \backslash U_\alpha$. Clearly $\{y_\alpha : \alpha < \kappa\}$ is a discrete subspace of Ult$A$. □

The proof shows that $\mathrm{Depth}_{\mathrm{h}+}A$ and sA have the same attainment properties.

Since $\mathrm{Depth}_{\mathrm{h}-}A \leq \mathrm{Depth}_{\mathrm{H}-}A$, we have $\mathrm{Depth}_{\mathrm{h}-}A = \omega$ for any infinite BA A. Obviously ${}_{\mathrm{d}}\mathrm{Depth}_{\mathrm{S}+}A =$ Depth$A$ for any BA A.

The status of the derived function ${}_{\mathrm{d}}\mathrm{Depth}_{\mathrm{S}-}$ is not clear. Note that ${}_{\mathrm{d}}\mathrm{Depth}_{\mathrm{S}-}A = \omega$ for A the interval algebra on a cardinal κ: this follows upon considering the subalgebra of A generated by $\{\{\alpha\} : \alpha$ a non-limit ordinal $< \kappa\}$. Also, Koppelberg and Shelah have independently observed that if A is atomless and λ-saturated (in the model-theoretic sense), then ${}_{\mathrm{d}}\mathrm{Depth}_{\mathrm{S}-}A \geq \lambda$. To show this, suppose that B is a dense subalgebra of A. By induction choose elements $a_\alpha \in A$ and $b_\alpha \in B$ for $\alpha < \lambda$ so that $\alpha < \beta$ implies that $a_\alpha > a_\beta > b_\beta > 0$; the a_α's can be chosen by λ-saturation, and the b_α's by denseness. So the sequence $\langle b_\alpha : \alpha < \lambda \rangle$ shows that the depth of B is at least λ.

Next, clearly $[\omega, \mathrm{t}A) \subseteq \mathrm{Depth}_{\mathrm{Hs}}A$, by an argument very similar to that used for the function c. And, of course, $\mathrm{Depth}_{\mathrm{Hs}} \subseteq [\omega, \mathrm{t}A]$. Like for cellularity, there is a problem whether t$A \in \mathrm{Depth}_{\mathrm{Hs}}A$. This is trivially true if

tA is a successor cardinal or a limit cardinal of cofinality ω by Corollary 2.12 and Theorem 10.2 below. Now by Corollary 2.12, to find a BA A such that $tA \notin \mathrm{Depth}_{Hs}A$ is equivalent to finding A so that A has no homomorphic image B with $\mathrm{Depth}B = tA$ attained. It is possible to have an algebra A which has a homomorphic image B such that $\mathrm{Depth}B = tA$ but $\mathrm{Depth}B$ is not attained: take $A = B = \prod^{w}_{\alpha<\mathrm{cf}\kappa} A_\alpha$, where κ is a singular cardinal of cofinality $> \omega$, $\langle \mu_\alpha : \alpha < \mathrm{cf}\kappa \rangle$ is an increasing sequence of cardinals with sup κ, and A_α is the interval algebra on μ_α for each $\alpha < \mathrm{cf}\kappa$.

Problem 5. *Is* $tB \in \mathrm{Depth}_{Hs}B$ *for every infinite BA B?*

Clearly $\mathrm{Depth}_{Ss} = [\omega, \mathrm{Depth}A]$ for any infinite BA A.

We always have $\mathrm{Depth} \leq cA$ and the difference can be arbitrarily large, for example in a finite-cofinite algebra.

Next comes the relation Depth_{Sr}. It is easy to see that parts (i)-(iii) and (v) of Theorem 1.10 hold with cellularity replaced by depth. We do not know if Theorem 1.10 (iv) holds for depth:

Problem 6. *Is there an infinite cardinal* κ *and a BA A such that* $(\kappa, (2^\kappa)^+) \in \mathrm{Depth}_{Sr}A$, *while* $(\omega, (2^\kappa)^+) \notin \mathrm{Depth}_{Sr}A$?

Here is another problem, just typical of many:

Problem 7. *Does there exist a BA A such that* $(\omega_1, \omega_2) \in \mathrm{Depth}_{Sr}A$ *but* $(\omega, \omega_1) \notin \mathrm{Depth}_{Sr}A$?

The following theorem seems relevant to these problems:

Theorem 2.14. (GCH) *For every infinite cardinal* κ *there is an interval algebra A of power* κ^+ *such that every subalgebra of A of power* κ^+ *has depth* $\geq \kappa$.

PROOF. Let μ be minimal such that $\omega^\mu > \kappa$. Let L be the linearly ordered set ${}^\mu\mathbf{Q}$ under lexicographic order, where $\mathbf{Q}$ is the set of all rationals in [0,1). Set $D = \{f \in {}^\mu Q : \text{there is an } \alpha < \mu \text{ such that } f\beta = 0 \text{ for all } \beta > \alpha\}$. It is clear that $|D| \leq \kappa$ and D is dense in L in the sense that if $f, g \in L$ and $f < g$ then there is an $h \in D$ such that $f < h < g$. Let M be a subset of L of size κ^+ which includes D, and let A be the interval algebra on M. Suppose that B is a subalgebra of A of power κ^+. Let N be any subset of B with κ^+ elements; we shall first show that B includes a simply ordered subset of size κ^+; here we follow closely the proof of Theorem 15.22 in Part I of the handbook. For each $x \in N$ write

$$x = [a(1,x), b(1,x)) \cup \ldots \cup [a(m_x,x), b(m_x,x)),$$

where $a(1,x)$, $b(1,x)$, ..., $a(m_x,x)$, $b(m_x,x)$ are in $M \cup \{+\infty\}$ and $a(1,x) < b(1,x) < \cdots < a(m_x,x) < b(m_x,x)$. By going from x to $-x$ if necessary, we may assume that $a(1,x) = 0$ for all $x \in N$. We may assume that m_x does not depend on x, so we drop the subscript x. Now for each $x \in N$ we choose

$$c(1,x), \ldots, c(m,x), d(1,x), \ldots, d(m,x) \in D$$

so that $a(i,x) < c(i,x) < b(i,x) < d(i,x) < a(i+1,x)$ for all $i = 1, \ldots, m$ (omitting the term $a(i+1,x)$ for $i = m$, and also omitting $d(i,x)$ if $b(i,x) = \infty$. We may assume that the elements $c(i,x)$ and $d(i,x)$ do not actually depend on x; so we write simply c_i and d_i. Next, we may assume that for some k, $1 \leq k \leq m$, the elements $a(k,x)$, $x \in N$, are pairwise distinct (the argument below is similar if some elements $b(k,x)$, $x \in N$ are pairwise distinct). Note that $k > 1$. Now define a homomorphism f of B into the BA of all subsets of $L \cap [d_{k-1}, c_k)$ by setting $fu = u \cap [d_{k-1}, c_k)$ for all $u \in B$. Now by Theorem 15.18 of the BA handbook, Part I, there is an isomorphism from the range of f into B. But clearly f takes N onto a linearly ordered set of power κ^+, as desired.

That B has depth $\geq \kappa$ now follows from the Erdös-Rado theorem $(2^\lambda)^+ \to (\lambda^+)^2_\lambda$ in an obvious way. □

We note the following two obvious facts about $\mathrm{Depth}_{\mathrm{Hr}}$:

(1) If $(\kappa, \lambda) \in \mathrm{Depth}_{\mathrm{Hr}} A$, then $\kappa \leq \lambda \leq |A|$ and $\kappa \leq \mathrm{t}A$.

(2) If $\kappa \in [\omega, \mathrm{t}A)$ then there is a $\lambda \leq 2^\kappa$ such that $(\kappa, \lambda) \in \mathrm{Depth}_{\mathrm{Hr}} A$.

Also, the following examples are relevant: if A is the finite-cofinite algebra on κ, then $\mathrm{Depth}_{\mathrm{Hr}} A = \{(\omega, \lambda) : \omega \leq \lambda \leq \kappa\}$; if A is free on κ, then $\mathrm{Depth}_{\mathrm{Hr}} A = \{(\lambda, \mu) : \omega \leq \lambda \leq \mu \leq \kappa\}$. A Problem similar to Problem 6 for $\mathrm{Depth}_{\mathrm{Sr}}$ is open:

Problem 8. *Is there an infinite cardinal κ and a BA A such that $(\kappa, (2^\kappa)^+) \in \mathrm{Depth}_{\mathrm{Hr}} A$, while $(\omega, (2^\kappa)^+) \notin \mathrm{Depth}_{\mathrm{Hr}} A$?*

The analog of Problem 7 is consistently resolved by an example of Todorčević [87]: in L, for each regular non-weakly compact cardinal κ there is an interval algebra A of power κ with $\mathrm{Depth}' A = \kappa$ and such that each homomorphic image of power $\lambda < \kappa$ has depth λ.

Concerning special classes of BA's, first notice that Depth is the same as cellularity for complete BA's. It is possible to have $\mathrm{Depth} A < \mathrm{c}A$ for an interval algebra. For example, let τ be the order type of the real numbers, let L be an ordered set of type $0 + (\omega + \omega^*) \cdot \tau$, and let A be the interval algebra on L. It is easily seen that $\mathrm{Depth} A = \omega$ while $\mathrm{c}A = 2^\omega$.

3. TOPOLOGICAL DENSITY

We begin with some equivalents of this notion. A set X of non-zero elements of a BA A is said to be *centered* provided that it satisfies the finite intersection property. And A is called *κ-centered* if $A\backslash\{0\}$ is the union of κ centered sets.

Theorem 3.1. *For any infinite BA A, $\mathrm{d}A$ is equal to each of the following cardinals:*
$\min\{\kappa : A$ *is isomorphic to a subalgebra of* $\mathcal{P}\kappa\}$;
$\min\{\kappa : A$ *is* κ-*centered*$\}$;
$\min\{\kappa : A$ *is a union of* κ *proper filters*$\}$;
$\min\{\kappa : A$ *is a union of* κ *ultrafilters*$\}$.

PROOF. Call the five cardinals mentioned $\kappa_0, \ldots \kappa_4$ respectively, starting with $\mathrm{d}A$ itself. $\kappa_0 \leq \kappa_1$: Let g be an isomorphism of A into $\mathcal{P}\kappa$. For each $\alpha < \kappa$ let $F_\alpha = \{a \in A : \alpha \in ga\}$. Then, as is easily checked, F_α is an ultrafilter on A. Let $Y = \{F_\alpha : \alpha < \kappa\}$. We claim that Y is dense in $\mathrm{Ult}A$. For, let U be a non-empty open set in $\mathrm{Ult}A$. We may assume that $U = \mathcal{S}a$ for some $a \in A$. Thus $a \neq 0$, so choose $\alpha \in ga$. Then $a \in F_\alpha$, and so $F_\alpha \in Y \cap U$, as desired.

$\kappa_1 \leq \kappa_2$: Suppose that $A\backslash\{0\} = \bigcup_{\alpha<\lambda} X_\alpha$, where each X_α is centered. Extend each X_α to an ultrafilter F_α. For each $a \in A$ let $fa = \{\alpha < \lambda : a \in F_\alpha\}$. Clearly f is an isomorphism of A into $\mathcal{P}\lambda$, as desired.

Obviously $\kappa_2 \leq \kappa_3 \leq \kappa_4$.

$\kappa_4 \leq \kappa_0$: Let X be a dense subset of $\mathrm{Ult}A$. Then obviously $A\backslash\{0\} = \bigcup_{F\in X} F$, as desired. □

We begin the discussion of algebraic operations for d. If A is a subalgebra of B, then $\mathrm{d}A \leq \mathrm{d}B$, and the difference can be arbitrarily large. If A is a homomorphic image of B, then d can change either direction in going from B to A. Thus if B is is a large free BA and A is a countable homomorphic image of B, then d goes down. On the other hand, if $B = \mathcal{P}\omega$, and $A = \mathcal{P}\omega/\mathrm{fin}$, then $\mathrm{d}B = \omega$ while $\mathrm{d}A = 2^\omega$, since in A there is a disjoint set of size 2^ω. Next, $\mathrm{d}(A \times B) = \max(\mathrm{d}A, \mathrm{d}B)$ for infinite BA's A, B. To see this, note that $\geq$ is clear, since A and B are isomorphic to subalgebras of $A \times B$. For the other inequality, suppose that f (resp. g) is an isomorphism of A (resp. B) into $\mathcal{P}\kappa$ (resp. $\mathcal{P}\lambda$). Let

$$X = \{(0, \alpha) : \alpha < \kappa\} \cup \{(1, \alpha) : \alpha < \lambda\}.$$

We define h mapping $A \times B$ into $\mathcal{P}X$ by setting

$$h(a, b) = \{(0, \alpha) : \alpha \in fa\} \cup \{(1, \alpha) : \alpha \in gb\}$$

for all $(a, b) \in A \times B$. It is easily verified that h is an isomorphism of $A \times B$ into $\mathcal{P}X$, and this proves $\leq$. A similar idea works for products and weak products in general:

Theorem 3.2. *If $\langle A_i : i \in I \rangle$ is a system of non-trivial BA's, then*

$$d\left(\prod_{i\in I} A_i\right) = d\left(\prod_{i\in I}^{\mathrm{w}} A_i\right) = \sum_{i\in I} dA_i.$$

PROOF. First we work with the full product, showing that $\sum_{i\in I} \mathrm{d}A_i = \mathrm{d}(\prod_{i\in I} A_i)$. Clearly $\mathrm{d}A_i \leq \mathrm{d}(\prod_{i\in I} A_i)$ for each $i \in I$. Since $\prod_{i\in I} A_i$ has a system of $|I|$ disjoint elements, we also have $|I| \leq \mathrm{d}\left(\prod_{i\in I} A_i\right)$. This verifies $\leq$. The direction $\geq$ is proved as in the case of two factors, using the "disjoint union" of all of the algebras. And the argument for weak products is the same. □

Concerning ultraproducts, we do not know the full story. The following is fairly clear, though. Let $\langle A_i : i \in I \rangle$ be a system of infinite BA's, and F an ultrafilter on I. Then $\mathrm{d}(\prod_{i\in I} A_i/F) \leq |\prod_{i\in I} \mathrm{d}A_i/F|$. To see this, let f_i be an isomorphism of A_i into $\mathcal{P}(\mathrm{d}A_i)$ for each $i \in I$. Then the desired isomorphism g of $\prod_{i\in I} A_i/F$ into $\mathcal{P}(\prod_{i\in I} \mathrm{d}A_i/F)$ is given as follows: for any $x \in \prod_{i\in I} A_i$,

$$g(x/F) = \{y/F : y \in \prod_{i\in I} \mathrm{d}A_i \text{ and } \{i \in I : y_i \in f_i x_i\} \in F\}.$$

(This is easily verified.) So, we have the following problem:

Problem 9. *Does there exist a system $\langle A_i : i \in I \rangle$ of infinite BA's and an ultrafilter F on I such that $d(\prod_{i\in I} A_i/F) < |\prod_{i\in I} \mathrm{d}A_i/F|$?*

Clearly $\mathrm{d}(A \oplus B) = \max(\mathrm{d}A, \mathrm{d}B)$: if f is an isomorphism of A into $\mathcal{P}\kappa$ and g is an isomorphism of B into $\mathcal{P}\lambda$, then the following function clearly extends to an isomorphism of $A \oplus B$ into $\mathcal{P}(\kappa \times \lambda)$: for $a \in A$ and $b \in B$,

$$ha = fa \times \lambda \text{ and } hb = \kappa \times gb.$$

For free products of several algebras there is a much more general topological result. To prove it, we need the following lemma.

Lemma 3.3. *Let κ be an infinite cardinal. Then the product space ${}^{\kappa}{}^{2}\kappa$ has density $\leq \kappa$ (where κ has the discrete topology).*

PROOF. Let $D = \{f \in {}^{{}^\kappa 2}\kappa$:there is a finite subset M of κ such that for all $x, y \in {}^\kappa 2$, if $x \upharpoonright M = y \upharpoonright M$, then $fx = fy\}$. We show that $|D| \leq \kappa$. First,

$$D = \bigcup_{M \in [\kappa]^{<\omega}} \{f \in {}^{{}^\kappa 2}\kappa : \text{ for all } x, y \in {}^\kappa 2(\text{ if } x \upharpoonright M = y \upharpoonright M, \text{ then } fx = fy\}.$$

So, it suffices to take any finite $M \subseteq \kappa$ and show that $N \stackrel{\text{def}}{=} \{f \in {}^{{}^\kappa 2}\kappa$: for all $x, y \in {}^\kappa 2$, if $x \upharpoonright M = y \upharpoonright M$ then $fx = fy\}$ has power at most κ. For any $f \in N$, let $f' \in {}^{{}^M 2}\kappa$ be defined as follows: for any $x \in {}^M 2$, choose any $y \in {}^\kappa 2$ such that $x \subseteq y$ and let $f'x = fy$. Clearly the assignment $f \mapsto f'$ is one-one. So $|N| \leq \kappa$, as desired.

To show that D is dense in ${}^{{}^\kappa 2}\kappa$, let U be an open set in ${}^{{}^\kappa 2}\kappa$. We may assume that U has a very special form, namely that there is a finite subset F of ${}^\kappa 2$ and a function g mapping F into κ such that

$$U = \{f \in {}^{{}^\kappa 2}\kappa : g \subseteq f\}.$$

Now let G be a finite subset of κ such that $f \upharpoonright G \neq h \upharpoonright G$ for distinct $f, h \in F$. Define $k \in {}^{{}^\kappa 2}\kappa$ in the following way: for any $x \in {}^\kappa 2$, set $kx = gf$ if $x \upharpoonright G = f \upharpoonright G$ for some $f \in F$, otherwise let kx be 0. Clearly $k \in D \cap U$, as desired. □

Theorem 3.4. *Suppose that $\langle X_i : i \in I\rangle$ is a system of topological spaces each having at least two disjoint non-empty open sets. Then* $\mathrm{d}(\prod_{i \in I} X_i) = \max(\lambda, \sup_{i \in I} \mathrm{d}X_i)$, *where λ is the least cardinal such that $|I| \leq 2^\lambda$.*

PROOF. Clearly $\mathrm{d}X_i \leq \mathrm{d}(\prod_{i \in I} X_i)$ for each $i \in I$. Suppose that D is dense in $\prod_{i \in I} X_i$ but $2^{|D|} < |I|$. Let U_i^0 and U_i^1 disjoint non-empty open sets in X_i for all $i \in I$. For each $i \in I$ let

$$V_i = \{x \in \prod_{i \in I} X_i : x_i \in U_i^0\}.$$

Then our supposition implies that there are distinct $i, j \in I$ such that $V_i \cap D = V_j \cap D$. Let $W = \{x : x_i \in U_i^0 \text{ and } x_j \in U_j^1\}$. Choose $x \in W \cap D$. Then $x \in V_i$ but $x \notin V_j$, contradiction.

Up to this point we have proved the inequality $\geq$. Now for each $i \in I$, let D_i be dense in X_i with $|D_i| = \mathrm{d}X_i$. Set $\kappa = \max(\lambda, \sup_{i \in I}|D_i|)$. Then for each $i \in I$ there is a function f_i mapping κ onto D_i. Since $|I| \leq 2^\lambda$, we then get a continuous function from ${}^{{}^\kappa 2}\kappa$ onto $\prod_{i \in I} D_i$. Namely, let g be a one-one function from I into ${}^\kappa 2$. For each $x \in {}^{{}^\kappa 2}\kappa$ and each $i \in I$ let $(hx)_i = f_i x_{gi}$. Then h is the desired continuous function. To see that h is continuous, let U be basic open in $\prod_{i \in I} D_i$. Then there is a finite $F \subseteq I$ such that $\mathrm{pr}_i[U]$ is open in D_i for all $i \in F$ and $\mathrm{pr}_i[U] = D_i$ for all $i \in I \backslash F$. Let $L = \{l \in {}^{g[F]}\kappa : \forall i \in F(f_i l_{gi} \in \mathrm{pr}_i U)\}$. For each $l \in L$ the set $W_l \stackrel{\text{def}}{=} \{k \in {}^{\kappa 2}\kappa : l \subseteq k\}$ is open in ${}^{\kappa 2}\kappa$, and $h^{-1}[U] = \bigcup_{l \in L} W_l$. So h is continuous. Clearly h is maps onto $\prod_{i \in I} D_i$.

Now Lemma 3.3 yields the desired result. □

Corollary 3.5. *If A is a free BA on κ free generators, then* $\mathrm{d}A$ *is the smallest cardinal λ such that $\kappa \leq 2^\lambda$.* □

Next we treat the topological density of the union of a well-ordered chain. Suppose that $\langle B_\alpha : \alpha < \kappa \rangle$ is a strictly increasing sequence of BA's with union A. Then $\mathrm{d}A \leq \sum_{\alpha<\kappa} \mathrm{d}B_\alpha$. In fact, for each $\alpha < \kappa$ let X_α be a set of ultrafilters on A such that $\{F \cap B_\alpha : F \in X_\alpha\}$ is dense in $\mathrm{Ult}B_\alpha$. Then it is easy to see that $\bigcup_{\alpha<\kappa} X_\alpha$ is dense in $\mathrm{Ult}A$, as desired. In particular, if $\mathrm{d}B_\alpha \leq \lambda$ for each $\alpha < \kappa$, then $\mathrm{d}A \leq \kappa \cdot \lambda$. Also note that if κ and λ satisfy these conditions, then $\kappa \leq (2^\lambda)^+$; in fact, otherwise, with $\alpha = (2^\lambda)^+$ we would have $|B_\alpha| \geq (2^\lambda)^+$ since at least one new element is added at each stage, while $|B_\alpha| \leq 2^\lambda$ by Theorem 3.1. The following example is also relevant. Assume GCH, and let B be a free BA on free generators $\{x_\alpha : \alpha < \omega_2\}$, and for each $\alpha < \omega_2$ let A_α be the subalgebra of B generated by $\{x_\xi : \xi < \alpha\}$. Then $\mathrm{d}B = \omega_1$, while $\mathrm{d}A_\alpha = \omega$ for all $\alpha < \omega_2$. Also relevant is the following. Suppose that $\langle B_\alpha : \alpha < \omega_2 \rangle$ is a strictly increasing sequence of BA's such that $\mathrm{d}B_\alpha = \omega$ for all $\alpha < \omega_2$, and we assume CH. Then $\mathrm{d}(\bigcup_{\alpha<\omega_2} B_\alpha) \leq \omega_1$. This follows from a result in Shelah [80]. These observations still do not describe completely what happens to d under unions.

Concerning derived operations for topological density, we first mention:

Theorem 3.6. $\mathrm{d}_{\mathrm{H}+}A = \mathrm{d}_{\mathrm{h}+}A = \mathrm{hd}A$ *for any infinite BA A.*

PROOF. By definition, $\mathrm{hd}A = \mathrm{d}_{\mathrm{h}+}A$, and trivially $\mathrm{d}_{\mathrm{H}+}A \leq \mathrm{d}_{\mathrm{h}+}A$. Furthermore, if D is dense in $S \subseteq \mathrm{Ult}A$, then D is also dense in $\overline{S}$, so the other inequality is also clear. □

We have $\mathrm{d}_{\mathrm{H}-}A = \mathrm{d}_{\mathrm{h}-}A = \omega$ for infinite A, since by Sikorski's extension theorem there is a homomorphism of A onto an infinite subalgebra of $\mathcal{P}\omega$. Clearly $\mathrm{d}_{\mathrm{S}+}A = \mathrm{d}A$, $\mathrm{d}_{\mathrm{S}-}A = \omega$, and ${}_{\mathrm{d}}\mathrm{d}_{\mathrm{S}+}A = \mathrm{d}A$ for any infinite BA A. Furthermore, ${}_{\mathrm{d}}\mathrm{d}_{\mathrm{S}-}A = \mathrm{d}A$: if B is a dense subalgebra of A and f is an isomorphism of B into $\mathcal{P}\kappa$, then f can be extended to an isomorphism of A into $\mathcal{P}\kappa$, as desired.

The spectrum functions d_{Hs} and d_{Ss} have not been investigated. We mention the following specific problems:

Problem 10. *Is it true that always* $\mathrm{d}_{\mathrm{Hs}}A = [\omega, \mathrm{hd}A)$ *or* $\mathrm{d}_{\mathrm{Hs}}A = [\omega, \mathrm{hd}A]$ *for an infinite BA A?*

Problem 11. *Is it true that always* $d_{Ss}A = [\omega, dA]$ *for an infinite BA A?*

From Theorem 3.1 the inequality $cA \leq dA$ for every infinite BA A is obvious. The difference between cA and dA can be arbitrarily large, for example in free BA's. A further relationship between cA and dA is found in Shelah [80]: if $\lambda = \lambda^{<\kappa}$, B satisfies the κ-cc, $|B| = \lambda^+$, and κ is regular and uncountable, then $dB \leq \lambda$.

Not much is known about the relations d_{Sr} and d_{Hr}. It is relevant here that $dA \leq |A|$ always holds.

Turning to topological density for special classes of BA's, note first that if A is the finite-cofinite algebra on κ, then $d_{Sr}A = \{(\lambda, \lambda) : \omega \leq \lambda \leq \kappa\} = d_{Hr}A$. For interval algebras, we have one interesting inequality not true for BA's in general. It is actually true for linearly ordered spaces in general, and we give that general form, due to Kurepa [35]. This result has evidently been rediscovered by many people independently; see, e.g., Juhász [71].

Theorem 3.7. *If L is an infinite linearly ordered space, then* $dL \leq (cL)^+$.

PROOF. Assume the contrary. Set $\kappa = (cL)^+$. Let $\prec$ be a well-ordering of L. Now we set

$$N = \{p \in L : p \text{ is the } \prec\text{-least element of some neighborhood of } p\}.$$

Clearly N is dense in L. Hence $|N| > \kappa$. Now for each $p \in N$ let I_p be the union of all open intervals having p as their $\prec$-first element. Then, we claim,

(1) If $p \prec p'$, then $I_p \cap I_{p'} = 0$ or $I_{p'} \subset I_p$.

In fact, suppose $p \prec p'$ and $I_p \cap I_{p'} \neq 0$. This means that there exist an open interval U with $\prec$-first element p and an open interval U' with $\prec$-first element p' such that $U \cap U' \neq 0$; hence $U \cup U'$ is an open interval with both p and p' as members, and with p as $\prec$-first member. So, if V is any open interval with $\prec$-first element p', then $V \cup U \cup U'$ is an open interval with $\prec$-first element p, and hence $V \subseteq I_p$. This shows that $I_{p'} \subseteq I_p$. Since $p \in I_p \backslash I_{p'}$, (1) then follows.

Next, set

$$N_0 = \{p \in N : I_p \text{ is not contained in any other } I_{p'}\}.$$

Now $I_p \cap I_{p'} = 0$ for all distinct $p, p' \in N_0$, so $|N_0| < \kappa$. We continue inductively for all $\xi < \kappa$:

$$H_\xi = N \backslash \bigcup_{\eta < \xi} N_\eta;$$

$$N_\xi = \{p \in H_\xi : I_p \text{ is not contained in any other } I_{p'} \text{ for } p' \in H_\xi\}.$$

Note inductively that $|N_\xi| < \kappa$, and hence always $H_\xi \neq 0$. It follows that $|\bigcup_{\xi<\kappa} N_\xi| \leq \kappa$, so there is a $p \in N \backslash \bigcup_{\xi<\kappa} N_\xi$. Thus $p \in H_\xi$ for all $\xi < \kappa$. But then for each $\xi < \kappa$ there is a $p(\xi) \in N_\xi$ such that $I_p \subset I_{p(\xi)}$. In fact, there is a $q \in H_\xi$ such that $I_p \subset I_q$. Taking the smallest such q under $\prec$, we get the desired $p(\xi)$. Hence for all $\xi, \eta < \kappa$ we have $I_{p(\xi)} \subset I_{p(\eta)}$ or $I_{p(\eta)} \subset I_{p(\xi)}$. By the partition relation $\kappa \to (\kappa, \omega)^2$ we may assume that $p(\xi) \prec p(\eta)$ whenever $\xi < \eta < \kappa$, and hence the sequence $\langle I_{p(\xi)} : \xi < \kappa \rangle$ is strictly decreasing. For each $\xi < \kappa$ choose $x_\xi \in I_{p(\xi)} \backslash I_{p(\xi+1)}$. Let

$$K^l = \{x_\xi : x_\xi < p(\xi+1)\}, \quad K^r = \{x_\xi : x_\xi > p(\xi+1)\}.$$

Now if $\xi < \eta$ and $x_\eta, x_\xi \in K^l$, then $x_\xi < x_\eta$: otherwise, note that x_ξ is less than all members of $I_{p(\xi+1)}$; so $x_\eta \leq x_\xi < p(\eta+1)$ and $x_\eta, p(\eta+1) \in I_{p(\eta)}$, so $x_\xi \in I_{p(\eta)}$, contradiction. Similarly, if $\xi < \eta$ and $x_\eta, x_\xi \in K^r$, then $x_\xi > x_\eta$. But this means that there are κ disjoint open intervals, contradiction. □

The interval algebra of a Souslin line gives an example of an interval algebra A in which $\mathrm{c}A < \mathrm{d}A$; on the other hand, Martin's axiom implies that for an interval algebra, $\mathrm{c}A = \omega \Rightarrow \mathrm{d}A = \omega$ (see any set theory book). In general, the existence of an interval algebra A such that $\mathrm{c}A < \mathrm{d}A$ is connected with the generalized Souslin problem.

Since interval algebras are retractive, it follows that if B is a homomorphic image of an interval algebra A, then $\mathrm{d}B \leq \mathrm{d}A$. Hence $\mathrm{d}A = \mathrm{hd}A$ for every interval algebra A.

An example of a complete BA A for which $\mathrm{c}A < \mathrm{d}A$ can be obtained by taking A to be the completion of a large free BA.

4. π-WEIGHT

If A is a subalgebra of B, then πA can vary either way from πB; for clearly one can have $\pi A < \pi B$, and if we take $B = \mathcal{P}\omega$ and A the subalgebra of B generated by an independent subset of size 2^ω, then we have $\pi B = \omega$ and $\pi A = 2^\omega$. Similarly, if A is a homomorphic image of B: it is easy to get such A and B with $\pi A < \pi B$, and if we take $B = \mathcal{P}\omega$ and $A = B/\mathrm{Fin}$, then $\pi B = \omega$ while $\pi A = 2^\omega$ since A has a disjoint subset of size 2^ω. Turning to products, we have $\pi(\prod_{i\in I} A_i) = \max(|I|, \sup_{i\in I}\pi A_i)$ for any system $\langle A_i : i \in I\rangle$ of infinite BA's. For, $\geq$ is clear; now suppose D_i is a dense subset of A_i for each $i \in I$. Let

$$E = \{f \in \prod_{i\in I}(D_i \cup \{0\}) : fi \neq 0 \text{ for only finitely many } i \in I\}.$$

Clearly E is dense in $\prod_{i\in I} A_i$, and $|E| = \max(|I|, \sup_{i\in I}\pi A_i)$, as desired. The same argument shows that $\pi(\prod^{\mathrm{w}}_{i\in I} A_i) = \max(|I|, \sup_{i\in I}\pi A_i)$.

Turning to ultraproducts, it is clear that $\pi(\prod_{i\in I} A_i/F) \leq |\prod_{i\in I} \pi A_i/F|$; but we do not know whether equality holds:

Problem 12. *Is it true that always $\pi(\prod_{i\in I} A_i/F) = |\prod_{i\in I} \pi A_i/F|$?*

An easy argument shows that $\pi(\oplus_{i\in I}A_i) = \max(|I|, \sup_{i\in I}\pi A_i)$ for any system $\langle A_i : i \in I\rangle$ of Boolean algebras. In fact, if D_i is dense in A_i for each $i \in I$, then

$$E \stackrel{\text{def}}{=} \{d_0 \cdot \cdots \cdot d_{n-1} : \exists \text{ distinct } i_0, \ldots, i_{n-1} \in I \text{ such that } \forall j < n(d_j \in D_{i_j})\}$$

is clearly dense in $\oplus_{i\in I}A_i$, and it has the indicated cardinality. On the other hand, suppose X is dense in $\oplus_{i\in I}A_i$. We may assume that each element of X is a product of members of $\bigcup_{i\in I} A_i$, with distinct factors coming from distinct A_i's. For each $i \in I$ let $Y_i = \{x \in X : x \leq a \text{ for some } a \in A_i\}$. For each $x \in Y_i$, let x_i^+ be obtained from x by replacing each factor of x which is not in A_i by 1. Clearly then $\{x_i^+ : x \in Y_i\}$ is still dense in $\oplus_{i\in I}A_i$, so $\pi A_i \leq |\{x_i^+ : x \in Y_i\}| \leq |Y_i| \leq |X|$. It is also clear that $|I| \leq |X|$; so $|X| \geq \max(|I|, \sup_{i\in I}\pi A_i)$, as desired.

We turn to the discussion of unions. The following theorem describes what happens.

Theorem 4.1. *Suppose that $\langle A_\alpha : \alpha < \kappa\rangle$ is a strictly increasing sequence of BA's, with union B, where κ is regular. Let $\lambda = \sup_{\alpha<\kappa}\pi A_\alpha$. Then $\kappa \leq 2^\lambda$, $\pi B \leq \max(\kappa, \lambda)$, and $\pi B \leq \lambda^+$.*

PROOF. We may assume that at limit α, A_α is the union of preceding algebras. Let $\mu = (2^\lambda)^+$. Suppose that $\kappa \geq \mu$, and let C be B if $\kappa = \mu$, and let $C = A_\mu$ otherwise. Let $S = \{\alpha < \mu : \text{cf}\alpha = \lambda^+\}$. Thus S is stationary in μ. For each $\alpha \in S$, A_α has a dense subset D_α of size $\leq \lambda$. Since $\text{cf}\alpha = \lambda^+$, it follows that there is an $f\alpha < \alpha$ such that $D_\alpha \subseteq A_{f\alpha}$. Now f is regressive on S, so f is constant on a stationary subset S' of S. Let β be the constant value of f on S'. Then D_β is dense in C. But $|C| \geq (2^\lambda)^+$, contradiction.

This shows that $\kappa \leq 2^\lambda$. Now as in the discussion of unions for topological density, we have $\pi B \leq \max(\kappa, \lambda)$. Supposing that $\pi B > \lambda^+$, this means that $\kappa > \lambda^+$. We can use the argument of the preceding paragraph to arrive at the contradiction that B has a dense subset of power at most λ. □

Note that the upper bound λ^+ mentioned in Theorem 4.1 can be attained: use a free algebra, as in the discussion of unions for topological density.

Turning to the functions derived from π, we first want to show that $\pi_{\text{H}+} = \pi_{\text{h}+} = \text{hd}$ (hereditary density). To do this, we need to discuss several other matters first; these will be useful later too. First note that π has a clear meaning for an arbitrary topological space X: namely, πX is the smallest cardinality of a family $\mathcal{O}$ of open sets such that for every non-empty open set U there is a $V \in \mathcal{O}$ such that $0 \neq V \subseteq U$. Such a family $\mathcal{O}$ is called a *π-base* for X. It is clear that for any space X, $\text{d}X \leq \pi X$.

Next, we call a sequence $\langle x_\xi : \xi < \kappa \rangle$ of elements of a space X *left-separated* provided that for every $\xi < \kappa$ there is an open set U in X such that $U \cap \{x_\eta : \eta < \kappa\} = \{x_\eta : \xi \leq \eta\}$. We now need the following important fact relating this notion to the function hd:

Theorem 4.2. *For any infinite Hausdorff space X, $\text{hd}X$ is the supremum of all cardinals κ such that there is a left-separated sequence in X of type κ.*

PROOF. If $\langle x_\xi : \xi < \kappa \rangle$ is a left-separated sequence in X and κ is infinite and regular, then clearly the density of $\{x_\xi : \xi < \kappa\}$ is κ. Hence the inequality $\geq$ holds. Now suppose that Y is a subspace of X, and set $\text{d}Y = \kappa$. We construct a left-separated sequence $\langle x_\xi : \xi < \kappa \rangle$ as follows: having constructed x_η for all $\eta < \xi$, where $\xi < \kappa$, it follows that $\{x_\eta : \eta < \xi\}$ is not dense in Y, and so we can choose $x_\xi \in Y \setminus \overline{\{x_\eta : \eta < \xi\}}$. This proves the other inequality. □

Note that the proof of Theorem 4.2 shows that if $\text{hd}X$ is attained, then it is also attained in the left-separated sense.

The essential fact used to prove that $\pi_{\text{H}+} = \text{hd}$ is as follows; we follow the proof of van Douwen [89], 8.1.

Lemma 4.3. *Let A be an infinite BA. Then there exists a sequence $\langle x_\xi : \xi < \pi A \rangle$ such that $\{x_\xi : \xi < \pi A\}$ is dense in A, and for each $\xi < \pi A$ and each finite subset G of $(\xi, \pi A)$ we have $x_\xi \cdot \prod_{\eta \in G} -x_\eta \neq 0$.*

PROOF. For brevity let $\pi = \pi A$. The major part of the proof consists in proving

(1) There is a sequence $\langle a_\xi : \xi < \pi\rangle$ of non-zero members of A such that $\{a_\xi : \xi < \pi\}$ is dense in A and for each $\eta < \pi$, $|\{\xi < \eta : a_\xi \cdot a_\eta \neq 0\}| < \pi(A \upharpoonright a_\eta)$.

To prove this, call an element $b \in A$ *π-homogeneous* provided that $\pi(A \upharpoonright c) = \pi(A \upharpoonright b)$ for every non-zero $c \leq b$. Clearly the collection of all π-homogeneous elements of A is dense in A. Let $\mathcal{A}$ be a maximal disjoint collection of π-homogeneous elements of A. Let $\kappa = |\mathcal{A}|$; then $\kappa \leq \mathrm{c}A \leq \pi A$. For each $b \in \mathcal{A}$ let M_b be a dense subset of $A \upharpoonright b$ of cardinality $\pi(A \upharpoonright b)$. Then $\bigcup\{M_b : b \in \mathcal{A}\}$ is dense in A. Now let $\langle N_b : b \in \mathcal{A}\rangle$ be a partition of π into disjoint subsets of power π. For each $b \in \mathcal{A}$ let f_b be a one-one function from M_b onto a subset of N_b of order type $\pi(A \upharpoonright b)$. Now for each $\xi < \pi$, let

$$a_\xi = \begin{cases} 0, & \text{if } \xi \notin \bigcup_{b\in\mathcal{A}} \mathrm{ran}(f_b); \\ f^{-1}\xi, & \text{if } \xi \in \mathrm{ran}(f_b),\ b \in \mathcal{A}. \end{cases}$$

Suppose that $\eta < \pi$ and $a_\eta \neq 0$. Say $\eta \in \mathrm{ran}(f_b)$. Then

$$|\{\xi < \eta : a_\xi \cdot a_\eta \neq 0\}| \leq |\{\xi \in \mathrm{ran} f_b : \xi < \eta\}| < \pi(A \upharpoonright b).$$

Thus we have (1), except that some of the a_η's are zero. If we renumerate the non-zero a_η's in increasing order of their indices, we really get (1).

Now we construct a sequence $\langle b_\alpha : \alpha < \pi\rangle$ of non-zero elements of A so that the following conditions hold:

(2_α) $b_\alpha \leq a_\alpha$ for all $\alpha < \pi$,

and

(3_α) for all $\xi < \alpha$ and every finite $F \subseteq (\xi, \alpha]$ we have $b_\xi \cdot \prod_{\eta\in F} -b_\eta \neq 0$ for all $\alpha < \pi$.

Assume that $\beta < \pi$, and b_α has been constructed for all $\alpha < \beta$ so that (2_α) and (3_α) hold. Then by (1) and (2_α) for all $\alpha < \beta$ we see that the set $\Gamma \stackrel{\text{def}}{=} \{\alpha < \beta : b_\alpha \cdot a_\beta \neq 0\}$ has power $< \pi(A \upharpoonright a_\beta)$. Hence there is a non-zero b_β in A such that $b_\beta \leq a_\beta$ and for all $\phi \in \Gamma$ and all finite $G \subseteq \Gamma$, if $b_\phi \cdot \prod_{\gamma\in G} -b_\gamma \neq 0$, then $b_\phi \cdot \prod_{\gamma\in G} -b_\gamma \not\leq b_\beta$. Thus (2_β) and (3_β) hold, and the construction is complete.

It is clear from (2_α) and (3_α) that $\langle b_\alpha : \alpha < \pi\rangle$ is the desired dense sequence. □

The following theorem is due to Shapirovskiĭ.

Theorem 4.4. *$\pi_{\mathrm{H}+}A = \pi_{\mathrm{h}+}A = \mathrm{hd}A$ for any infinite BA A.*

PROOF. It is obvious that $\pi_{\mathrm{H}+}A \leq \pi_{\mathrm{h}+}A$. Now if $\mathcal{O}$ is a π-base for $Y \subseteq \mathrm{Ult}A$ with $|\mathcal{O}| = \pi Y$, without loss of generality $\mathcal{O}$ has the form $\{Sa \cap Y : a \in \mathcal{A}\}$

for some $\mathcal{A} \subseteq A$. Let $fx = \mathcal{S}x \cap Y$ for any $x \in A$. Then f is a homomorphism onto some algebra B of subsets of Y, and $\mathcal{O}$ is dense in B. This shows that $\pi_{\mathrm{h}+}A \leq \pi_{\mathrm{H}+}A$. It is also trivial that $\mathrm{hd}A \leq \pi_{\mathrm{H}+}A$: if B is a homomorphic image of A, then $\mathrm{d}B \leq \pi B$. It remains just to show that $\pi_{\mathrm{H}+}A \leq \mathrm{hd}A$. Suppose that f is a homomorphism of A onto B, where B is infinite. Apply 4.3 to B to get a system $\langle b_\xi : \xi < \pi B\rangle$ of elements of B such that for any $\xi < \pi B$ and any finite subset G of $(\xi, \pi B)$ we have $b_\xi \cdot \prod_{\eta\in G} -b_\eta \neq 0$. For each $\xi < \pi B$ choose a_ξ so that $fa_\xi = b_\xi$, and let F_ξ be an ultrafilter containing all of the elements $a_\xi \cdot \prod_{\eta\in G} -a_\eta$ for G a finite subset of $(\xi, \pi B)$. We claim that $\langle F_\xi : \xi < \pi B\rangle$ is a left-separated sequence in Ult$A$, so that 4.2 yields the desired conclusion. In fact, for any $\xi < \pi B$ we have

$$\{F_\rho : \rho < \pi B\} \cap \bigcup_{\xi\leq\eta} \mathcal{S}a_\eta = \{F_\rho : \rho \leq \eta\}. \qquad \square$$

The proof of Theorem 4.4 shows that $\pi_{\mathrm{H}+}$ and $\pi_{\mathrm{h}+}$ have the same attainment properties; also, if $\pi_{\mathrm{H}+}$ is attained, then hd is attained in the left-separated sense, and if hd is attained, then $\pi_{\mathrm{H}+}$ is attained.

The cardinal function $\pi_{\mathrm{S}+}$ is of some interest, since it does not coincide with any of our standard ones. Obviously $\pi A \leq \pi_{\mathrm{S}+}A$ for any infinite BA A. Moreover, $\pi_{\mathrm{S}+}A \leq \pi_{\mathrm{H}+}A$; this follows from the following fact: for every subalgebra B of A there is a homomorphic image C of A such that $\pi B = \pi C$. To see this, by the Sikorski extension theorem extend the identity function from B into $\overline{B}$ to a homomorphism from A onto a subalgebra C of $\overline{B}$. Since $B \subseteq C \subseteq \overline{B}$, it is clear that $\pi B = \pi C$. Thus we have shown that $\pi A \leq \pi_{\mathrm{S}+}A \leq \pi_{\mathrm{H}+}A$ for any infinite BA A. It is possible to have $\pi A < \pi_{\mathrm{S}+}A$: let $A = \mathcal{P}\kappa$—then $\pi A = \kappa$, while $\pi_{\mathrm{S}+}A = 2^\kappa$, since A has a free subalgebra B of power 2^κ, and clearly $\pi B = 2^\kappa$. It is more difficult to come up with an example of an algebra where the other inequality is proper (this example is due to Monk):

EXAMPLE 4.5. *There is an infinite BA A such that $\pi_{\mathrm{S}+}A < \pi_{\mathrm{H}+}A$*

To see this, let B be the interval algebra on the real numbers, and let $A = B \oplus B$. Now, we claim, $\pi_{\mathrm{S}+}A = \omega$, while $\pi_{\mathrm{H}+}A = 2^\omega$. The latter equality holds since $\mathrm{s}A = 2^\omega$ by the argument in the proof of 1.8. We repeat this argument here in different terminology. For each real number r let $c_r = b_r \cdot b'_r$, where $b_r = [-\infty, r)$ (as a member of the first factor of $B \oplus B$) and $b'_r = [r, \infty)$ (as a member of the second factor of $B \oplus B$). Note that we have adjoined $-\infty$ as a member of $\mathbb{R}$ in order to fulfill the requirement for interval algebras that the ordered set in question always has a first element. It is easily checked that $\langle c_r : r \in \mathbb{R}\rangle$ forms an ideal independent system of elements of A, as desired. Clearly $\mathrm{s}A \leq \mathrm{hd}A$ by 4.2; then use 4.4. To prove that $\pi_{\mathrm{S}+}A = \omega$, we proceed as follows. Let C be any subalgebra of A. We want to show that $\pi C = \omega$. Now each element c of C can be written in the form

$$\sum_{i<m(c)} x_{0ic} \times x_{1ic},$$

where x_{0ic} and x_{1ic} are half-open intervals in B. Let $T = \{(m, r) : m \in \omega \backslash \{0\}$ and $r \in {}^m\mathbf{Q}\}$. An element (m, r) of T is a *frame* for $c \in C$ provided that $m(c) = m$ and $r_i \in x_{0ic}$ for all $i < m$. For each $(m, r) \in T$, let D_{mr} be the set of all $c \in C$ with frame (m, r). Since C is the union of all of the sets D_{mr}, it suffices to take an arbitrary $(m, r) \in T$ and find a countable dense subset of D_{rm}.

We do a similar thing for the x_{1ic}'s. Namely, for each $c \in D_{rm}$ write

$$x_{1ic} = \bigcup_{j<n(i,c)} y_{ijc},$$

where each y_{ijc} is a half-open interval of reals. Let $U = \{(n, s) : n \in {}^m(\omega\backslash\{0\})$ and $s \in \prod_{i<m} {}^{n(i)}\mathbf{Q}\}$. A member (n, s) of U is a *subframe* of $c \in D_{rm}$ provided that $n(i, c) = n(i)$ and $s_i(j) \in y_{ijc}$ for all $i < m$. For each $(n, s) \in U$, let E_{ns} be the set of all $c \in D_{rm}$ such that (n, s) is a subframe of c. Since D_{rm} is the union of all of these sets E_{ns}, it suffices now to take any $(n, s) \in U$ and find a countable dense subset of E_{ns}.

E_{ns} is clearly closed under $\cdot$. Hence it suffices to prove the following:

(1) If $i < m$ and $j < n(i)$, then there is a countable $N \subseteq E_{ns}$ such that for every $d \in E_{ns}$ there is a $c \in N$ such that $x_{0ic} \cdot y_{ijc} \leq x_{0id} \cdot y_{ijd}$.

To prove (1), let N_0 be a countable set of elements of E_{ns} such that the left end-points of the elements x_{oid} for $d \in N_0$ are cofinal in the set of left end-points of all elements x_{oic} for $c \in E_{ns}$. Similarly let N_1 do the same thing for coinitiality and right end-points, N_2 the same thing for cofinality, left end-points, and elements y_{ijc}, and N_3 the same thing for coinitiality, right end-points, and elements y_{ijc}. Finally, let

$$N = \{a_0 \cdot a_1 \cdot a_2 \cdot a_3 : a_i \in N_i \text{ for all } i < 4\}.$$

To show that (1) holds, let $d \in E_{ns}$. Then let $z(0)$ be an element of N such that the left end-point of $x_{0iz(0)}$ is $\geq$ the left end-point of x_{0id}. Similarly choose $z(1)$ for $\leq$ and the right end-point, $z(2)$ for $\geq$ and the left end-point of y_{ijd}, and $z(3)$ for $\leq$ and the right end-point of y_{ijd}. Then let $c = z(0) \cdot z(1) \cdot z(2) \cdot z(3)$. Clearly $c \in N$ and $x_{0ic} \cdot y_{ijc} \leq x_{0id} \cdot y_{ijd}$, as desired. □

The following problem is of some interest:

Problem 13. *Is* $\pi_{S+}A$ *always attained?*

Clearly $\pi_{S-}A = \pi_{H-}A = \pi_{h-}A = \omega$ for any infinite BA A. Furthermore, ${}_{d}\pi_{S+}A = {}_{d}\pi_{S-}A = \pi A$ for any infinite BA A. The spectra $\pi_{Hs}A$ and $\pi_{Ss}A$ have not been investigated, and we mention two obvious problems:

Problem 14. *Is it true that for every infinite BA A we have*

$$\pi_{Hs}A = \begin{cases} [\omega, \mathrm{hd}A], & \textit{if hdA is attained,} \\ [\omega, \mathrm{hd}A), & \textit{otherwise?} \end{cases}$$

Problem 15. *Is it true that for every infinite BA A we have*

$$\pi_{Ss}A = \begin{cases} [\omega, \pi_{S+}A], & \textit{if } \pi_{S+}A \textit{ is attained,} \\ [\omega, \pi_{S+}A), & \textit{otherwise?} \end{cases}$$

We have already observed that $\mathrm{d} \leq \pi$; the difference is small, though, since $\mathrm{d}A \leq \pi A \leq |A| \leq 2^{\mathrm{d}A}$ for any infinite BA A. Also worth noting is that under MA+$\neg$CH, if $\mathrm{c}A = \omega$ and $\pi A < 2^{\omega}$ then $\mathrm{d}A = \omega$; see Hajnal, Juhász [71].

We have not investigated the relations π_{Sr} and π_{Hr}.

About π for special classes of BAs, note that $\pi A = \mathrm{d}A$ for any interval algebra A; in fact, πA is also equal to $\mathrm{hd}A$. To see this, note that $\mathrm{d}A = \mathrm{hd}A$ for A an interval algebra, since any interval algebra is retractive; then $\pi A = \mathrm{d}A$ by the above inequalities. For A atomic, clearly πA is the number of atoms of A. Also note that $\pi_{S+}A = |A|$ for A complete, and $\pi_{S+}A = \mathrm{hd}A$ for A retractive.

If A is the completion of the free BA on ω_1 free generators, then $\mathrm{d}A < \pi A$: clearly $\pi A = \omega_1$, and Sikorski's extension criterion can be used to show that $\mathrm{d}A = \omega$.

5. LENGTH

Recall that Length$A$ is the sup of cardinalities of subsets of A which are simply ordered by the Boolean ordering. For references see the beginning of section 2. The analysis of Length is similar to that for Depth; many of the proofs are similar, but there are some differences. To take care of the first problem, attainment of Length, we need two small lemmas about orderings:

Lemma 5.1. *Let L be a linear ordering of regular cardinality λ which has no strictly increasing or strictly decreasing sequences of length λ. Then there exist $a < b$ in L such that $|[a,b)| = \lambda$.*

PROOF. Let $\langle a_\xi : \xi < \alpha\rangle$ and $\langle b_\xi : \xi < \beta\rangle$ be coinitial strictly decreasing and cofinal strictly increasing sequences in L, respectively. Then L, except for its greatest element, if it has such, is the union of all of the intervals $[a_\xi, b_\eta)$, and $\alpha < \lambda$ and $\beta < \lambda$, so the conclusion is clear. □

Lemma 5.2. *Let L be a linear ordering with first element 0, and with cardinality κ^+, where κ is infinite. Then there exist $a < b$ in L such that $|[a,b)| \geq \kappa$ and $|L\backslash[a,b)| \geq \kappa$.*

PROOF. Suppose not. Then clearly

(1) in there is no strictly increasing or strictly decreasing sequence of length κ^+.

Define by induction a sequence $\langle[a_\xi, b_\xi) : \xi < \alpha\rangle$ of half-open intervals in L such that $[a_\eta, b_\eta) \subset [a_\xi, b_\xi)$ for $\xi < \eta$, $|[a_\xi, b_\xi)| = \kappa^+$, and $|L\backslash[a_\xi, b_\xi)| < \kappa$ for all $\xi < \alpha$, continuing as long as possible. How long is this? Well, if $[a_\xi, b_\xi)$ has been defined, then $[a_{\xi+1}, b_{\xi+1})$ can clearly be defined. Suppose that $[a_\xi, b_\xi)$ has been defined for all $\xi < \beta$, where β is a limit ordinal $< \kappa^+$. Then

$$|L\backslash \bigcap_{\xi<\beta}[a_\xi, b_\xi)| = |\bigcup_{\xi<\beta} L\backslash[a_\xi, b_\xi)| < \kappa^+,$$

so by (1) and Lemma 5.1 applied to $\bigcap_{\xi<\beta}[a_\xi, b_\xi)$, the interval $[a_\beta, b_\beta)$ can be defined. Thus $\alpha \geq \kappa^+$. Now $a_\xi \leq a_\eta$ and $b_\xi \geq b_\eta$ for $\xi < \eta$, so one of $\{a_\xi : \xi < \kappa^+\}$ and $\{b_\xi : \xi < \kappa^+\}$ contains a suborder of L of size κ^+. This contradicts (1). □

Theorem 5.3. *If* cf(Length$A$) $= \omega$, *then* Length$A$ *is attained.*

PROOF. The proof should be fairly clear, following the lines of the proof of 2.2. Some modifications: a is an ∞-*element* provided that for each $i \in \omega$, some ordering of size λ_i is embeddable in $A \upharpoonright a$. When constructing a_i, Lemma 5.2

is used to obtain elements c, d such that $b = c + d$, $c \cdot d = 0$, and both $A \upharpoonright c$ and $A \upharpoonright d$ contain strictly increasing chains of length λ_i; then the new (*) is applied. □

The analog of 2.3 for Length does not hold. For example, if A is any denumerable BA, then ${}^{\omega}A$ has length 2^{ω}. This is because $\mathcal{P}\mathbf{Q}$ can be embedded in ${}^{\omega}A$, and $\mathbf{R}$ can be embedded in $\mathcal{P}\mathbf{Q}$: for each $r \in \mathbf{R}$, let $fr = \{q \in \mathbf{Q} : q < r\}$. To generalize this example, let us call a subset D of a linear order L *weakly dense in* L provided that if $a, b \in L$ and $a < b$, then there is a $d \in D$ such that $a \leq d \leq b$. Now for any infinite cardinal κ let $\mathrm{Ded}\kappa = \sup\{\lambda$: there is an ordering of size λ with a weakly dense subset of size $\kappa\}$. The following theorem from Kurepa [57] shows the connection of this notion to length in $\mathcal{P}\kappa$:

Theorem 5.4. *Let κ and λ be cardinals such that $\omega \leq \kappa \leq \lambda$. Then the follolwing two conditions are equivalent:*

(i) There is an ordering L of size λ with a weakly dense subset of size κ.

(ii) In $\mathcal{P}\kappa$ there is a chain of size λ.

Proof. $(i) \Rightarrow (ii)$. We may assume that $\kappa < \lambda$. Let D be weakly dense in L, with $|D| = \kappa$. Thus $|L \backslash D| = \lambda$. Let f be a one-one function from κ onto D. For each $a \in L\backslash D$ let $ga = \{\alpha < \kappa : f\alpha < a\}$. Clearly $a < b$ implies that $ga \subseteq gb$. Suppose $a < b$ with $a, b \in L\backslash D$; choose $x \in D$ so that $a \leq x \leq b$. Hence $a < x < b$, and so $f^{-1}x \in gb\backslash ga$ and $ga \neq gb$, as desired.

$(ii) \Rightarrow (i)$. Let L be a chain in $\mathcal{P}\kappa$ of size λ. For each $\alpha < \kappa$ let $x_\alpha = \bigcup\{a \in L : \alpha \notin a\}$. For any $a \in L$ and $\alpha < \kappa$ we clearly have $a \subseteq x_\alpha$ or $x_\alpha \subseteq a$; hence we may assume that $\{x_\alpha : \alpha < \kappa\} \subseteq L$. Let D be a subset of L of size κ such that $\{x_\alpha : \alpha < \kappa\} \subseteq D$. Now suppose that $a, b \in L$ and $a \subset b$. Choose $\alpha \in b\backslash a$. Then $a \subseteq x_\alpha$; and if $c \in L$ and $\alpha \notin c$, then $c \subseteq b$; so $x_\alpha \subseteq b$, as desired. □

Because of this theorem, about all that we can say about the length of products is this:

$$\max(\mathrm{Ded}|I|, \sup_{i\in I}\mathrm{Length}A_i) \leq \mathrm{Length}\left(\prod_{i\in I} A_i\right) \leq \prod_{i\in I}\mathrm{Length}A_i.$$

Shelah [87] has shown that $\mathrm{Length}(\prod_{i\in I} A_i)$ cannot be calculated purely from $|I|$ and $\langle \mathrm{Length}A_i : i \in I\rangle$.

For weak products, we clearly have the following analogs of 2.6 and 2.7:

Theorem 5.5. *Let $\kappa = \sup_{i\in I}\mathrm{Length}\ A_i$, and suppose that $\mathrm{cf}\kappa > \omega$. Then the following conditions are equivalent:*

(i) $\prod^{\mathrm{w}}_{i\in I} A_i$ has no chain of size κ.

(ii) For all $i \in I$, A_i has no chain of size κ.

PROOF. For the non-trivial direction $(ii) \to (i)$, suppose that X is a chain in $\prod^{w}_{i\in I} A_i$. Wlog assume that for each $x \in X$, the set $M_x \stackrel{\text{def}}{=} \{i \in I : x_i \neq 0\}$ is finite. Define $x \equiv y$ iff $M_x = M_y$. Then it is easy to see that $\equiv$ is a convex equivalence relation on X; there is an order induced on $X/\equiv$, and clearly that order is an interval of the ordered set ω. It follows from $\mathrm{cf}\kappa > \omega$ that some equivalence class has cardinality κ. Then the argument can go as in the case of depth. □

Corollary 5.6. $\mathrm{Length}(\prod^{w}_{i\in I} A_i) = \sup_{i\in I}\mathrm{Length}A_i$. □

By 5.5 we see that 5.3 is best possible: if κ is a limit cardinal with $\mathrm{cf}\kappa > \omega$, then it is easy to construct a weak product B such that $\mathrm{Length}B = \kappa$ but length is not attained in B.

If A is a subalgebra of B, then $\mathrm{Length}A \leq \mathrm{Length}B$, and the difference can be arbitrarily large. If A is a homomorphic image of B, then length can vary either way from B to A again, see the argument for cellularity. Since an ultraproduct of chains is a chain, the question of length of ultraproducts reduces to some extent to the question of cardinality of ultraproducts: if $\kappa_i < \mathrm{length}'A_i$ for all $i \in I$, then $\left|\prod_{i\in I} \kappa_i/F\right| \leq \left|\mathrm{length}\left(\prod_{i\in I} A_i/F\right)\right|$.

For free products, we have $\mathrm{Length}(\oplus_{i\in I}A_i) = \sup_{i\in I}\mathrm{Length}A_i$; this result of Grätzer and Lakser was considerably generalized by McKenzie and Monk; but in any case the proof is too lengthy to include here. There are some problems connected with free products and length; for details, see McKenzie, Monk [82]; but we mention this problem:

Problem 16. *Let A be the interval algebra of a Souslin line, B a BA with length $\aleph_{\omega_1}$ not attained. Is* $\mathrm{Length}(A\oplus B)$ *attained?*

Length is an ordinary sup-function, so Theorem 1.1 applies.

We turn to derived functions for length. The function $\mathrm{Length}_{\mathrm{H}+}A$ seems to be new. Note just that $\mathrm{t}A = \mathrm{Depth}_{\mathrm{H}+}A \leq \mathrm{Length}_{\mathrm{H}+}A$, using 2.11. It is possible to have $\mathrm{t}A < \mathrm{Length}_{\mathrm{H}+}A$; this is true when A is the interval algebra on $\mathbf{R}$, since $\mathrm{t}A = \omega$, while obviously $\mathrm{Length}_{\mathrm{H}+}A = 2^{\omega}$. To see that $\mathrm{t}A = \omega$, one can use 1.2 and 1.7 plus the following simple relationship which we need later anyway:

Theorem 5.7. $\mathrm{t}A \leq \mathrm{s}A$.

PROOF. We use 2.10: let $\langle F_\xi : \xi < \alpha\rangle$ be a free sequence in $\mathrm{Ult}A$. Set $X = \{F_\xi : \xi < \alpha\}$. For each $\xi < \alpha$,

$$(\mathrm{Ult}A\backslash\overline{\{F_\eta : \eta < \xi\}}) \cap (\mathrm{Ult}A\backslash\overline{\{F_\eta : \xi + 1 \leq \eta < \alpha\}})$$

is an open set whose intersection with X is $\{F_\xi\}$, as desired. □

We mention one problem concerning (implicitly) $\text{Length}_{\text{H}+}$:

Problem 17. *Is it consistent to have a BA A with the property that* $\omega < \text{Length}A < |A|$, *while A has no homomorphic image of power* $< |A|$*?*

The function $\text{Length}_{\text{H}-}$ is also new. Note that $\omega \leq \text{Length}_{\text{H}-}A \leq 2^\omega$, by an easy argument using the Sikorski extension theorem. It is obviously possible to have $\omega = \text{Length}_{\text{H}-}A$.

Problem 18. *Is always* $\text{Length}_{\text{H}-}A = \text{Card}_{\text{H}-}A$?

Obviously $\text{Length}_{\text{S}+}A = \text{Length}A$ and $\text{Length}_{\text{S}-}A = \omega$.

Problem 19. *Is always* $\text{Length}_{\text{h}-}A = \omega$?

Clearly $\text{Length}_{\text{h}+}A \geq \text{Depth}_{\text{h}+}A = \text{s}A$ by 2.13. And $\text{Length}_{\text{h}+}A \geq \text{Length}_{\text{H}+}A$; but it is possible to have $\text{Length}_{\text{h}+}A > \text{Length}_{\text{H}+}A$. This is true, for example, if A is the finite-cofinite algebra on an uncountable cardinal κ. For then $\text{Length}_{\text{h}+}A = \text{Ded}\kappa$, while $\text{Length}_{\text{H}+}A = \omega$. That $\text{Length}_{\text{h}+}A = \text{Ded}\kappa$ is seen like this: $\text{Ult}A$ has a discrete subspace S of size κ, and so Theorem 5.4 applies for the chains of subsets of S, since every subset is clopen.

Clearly ${}_{\text{d}}\text{Length}_{\text{S}+}A = \text{Length}A$.

Problem 20. *Is always* ${}_{\text{d}}\text{Length}_{\text{S}-}A = \omega$?

We have not investigated $\text{Length}_{\text{Hs}}$. Clearly $\text{Length}_{\text{Ss}}A = [\omega, \text{Length}A]$ for every infinite BA A.

Concerning the relationships of length to our previously treated functions, note that obviously $\text{Depth}A \leq \text{Length}A$ for any infinite BA A. Another clear relationship is $\text{Length}A \leq 2^{\text{Depth}A}$: if L is an ordered subset of A of power $(2^{\text{Depth}A})^+$, let $\prec$ be a well-ordering of L; then by the Erdös-Rado partition relation $(2^\kappa)^+ \rightarrow (\kappa^+)^2$ we get a well-ordered or inversely well-ordered subset of L of power $(\text{Depth}A)^+$, contradiction. Note that $\text{Length}A > \pi A$ for $A = \mathcal{P}\omega$; and $\text{c}A > \text{Length}A$ for A the finite-cofinite algebra on κ.

We have not investigated $\text{Length}_{\text{Sr}}$ and $\text{Length}_{\text{Hr}}$.

6. IRREDUNDANCE

Clearly $\text{Irr}A \le |A|$. If A is a subalgebra of B, then $\text{Irr}A \le \text{Irr}B$, and Irr can change to any extent from B to A (along with cardinality). The same is true for A a homomorphic image of B. Concerning the derived operations, we note just the obvious facts that $\text{Irr}_{\text{S}+}A = \text{Irr}A$, $\text{Irr}_{\text{S}-}A = \omega$, $\text{Irr}_{\text{h}-}A = \omega$, and ${}_{\text{d}}\text{Irr}_{\text{S}+}A = \text{Irr}A$. Obviously any chain is irredundant; so $\text{Length}A \le \text{Irr}A$. The difference can be large, e.g. in a free BA. By Theorem 4.25 of Part I of the BA handbook, $\pi A \le \text{Irr}A$. In particular, if $|A|$ is strong limit, then $|A| = \text{Irr}A$, since then $\pi A = |A|$.

These trivial facts give the immediate results about irredundance. Deeper facts about it are that it is consistent that there is a BA with irredundance less than cardinality, and it is also consistent that every uncountable BA has uncountable irredundance. The latter is an unpublished result of Todorčević, using an argument similar to his solution to the S-space problem. We shall spend the rest of this section proving the first fact, in the form that under CH there is a BA of power ω_1 with countable irredundance. We give two examples for this. The first example is $A \oplus A$, where A is the BA of closed-open subsets of a compact Kunen line. We say "a" since there are various Kunen lines, and we say "compact" since the standard Kunen lines are only locally compact. For the Kunen lines, see Juhász, Kunen, Rudin [76]. The second construction uses considerably less than CH, and can be found in Todorčević [86]. For a forcing construction of an uncountable BA with countable irredundance, see Bell, Ginsburg, Todorčević [82]. A generalization of the main results about irredundance (to other varieties of universal algebras) can be found in Heindorf [89].

The history of these results is complicated. I think that the first example of an uncountable BA with countable irredundance is due to Rubin [83] (the result was obtained several years before 1983). The papers with the constructions we give do not mention irredundance; their relevance for our purposes is due to a simple theorem of Heindorf [89]. So, modulo the simple theorem of Heindorf, the first example with irredundance different from cardinality is a Kunen line.

EXAMPLE 6.1. (CH) (*A compact Kunen line*). We construct a Boolean space making use of the topology on the real line; the resulting space is not linearly ordered, despite the name. We construct it by constructing a certain locally compact space, and then taking the one-point compactification to get the Boolean space we are interested in. Since we will be dealing with many topologies, we have to be precise about what we mean by a topology—for us, it is just the collection of all open sets. For any topology σ and any subset A of the space in question, $\bar{A}^\sigma$ denotes the closure of A with respect to the topology σ. Let $\langle x_\xi : \xi < \omega_1 \rangle$ be a one-one enumeration of $\mathbf{R}$. For each $\alpha \le \omega_1$ let $\mathbf{R}_\alpha = \{x_\xi : \xi < \alpha\}$. Let ρ be the usual topology on $\mathbf{R}$. Now we claim

(1) There is an enumeration $\langle S_\mu : \mu < \omega_1 \rangle$ of all of the countable subsets of $\mathbf{R} \times \mathbf{R}$ such that $S_\mu \subseteq \mathbf{R}_\mu \times \mathbf{R}_\mu$ for all $\mu < \omega_1$.

In fact, first let $\langle S'_\mu : \mu < \omega_1 \rangle$ be any old enumeration of the countable subsets of $\mathbf{R} \times \mathbf{R}$. We define $S_\mu = \mathbf{R}_\mu \times \mathbf{R}_\mu$ for all $\mu < \omega$. Now for $\omega \leq \mu < \omega_1$ let $S_\mu = S'_\nu$, where ν is minimum such that $S'_\nu \notin \{S_\eta : \eta < \mu\}$ and $S'_\nu \subseteq \mathbf{R}_\mu \times \mathbf{R}_\mu$. To see that this is the desired enumeration, suppose that S'_ν is not in the range of the function S, and choose ν minimum with this property. Then choose $\mu < \omega_1$ such that $S'_\rho \in \text{Rng}(S \restriction \mu)$ for each $\rho < \nu$ and $S'_\nu \subseteq \mathbf{R}_\mu \times \mathbf{R}_\mu$. Then the construction gives $S_\mu = S'_\nu$, contradiction.

Now we construct topologies τ_η for all $\eta \leq \omega_1$ so that the following conditions hold:

(2_η) τ_η is a topology on $\mathbf{R}_\eta$.

(3_η) $\tau_\xi = \tau_\eta \cap \mathcal{P}\mathbf{R}_\xi$ for $\xi < \eta$.

(4_η) $\tau_\eta \supseteq \{\mathbf{R}_\eta \cap U : U \in \rho\}$.

(5_η) If $\xi, \xi' < \eta$, $\mu < \xi$ or $\mu < \xi'$, and $(x_\xi, x_{\xi'}) \in \bar{S}^\rho_\mu$, then $(x_\xi, x_{\xi'}) \in \bar{S}^{\tau_\eta}_\mu$.

(6_η) τ_η is first-countable.

(7_η) τ_η is Hausdorff.

(8_η) In τ_η, the compact open sets form a base.

For $\beta \leq \omega$ let τ_β be the discrete topology on $\mathbf{R}_\beta$. Then the conditions (2_β)–(8_β) are clear; (5_β) holds since S_μ is finite under the indicated hypotheses.

Now assume that $\omega < \beta \leq \omega_1$ and τ_α has been constructed for all $\alpha < \beta$ so that $(2_\alpha) - (8_\alpha)$ hold. If β is a limit ordinal, let

$$\tau_\beta = \{U \subseteq \mathbf{R}_\beta : U \cap \mathbf{R}_\alpha \in \tau_\alpha \text{ for all } \alpha < \beta\}.$$

Then $(2_\beta) - (5_\beta)$ and (7_β) are clear. For (6_β), suppose that $\xi < \beta$; we want to find a countable neighborhood base for x_ξ. Let $\{U_n : n \in \omega\}$ be a countable neighborhood base for x_ξ in the topology $\tau_{\xi+1}$. If $V \in \tau_\beta$ and $x_\xi \in V$, then $V \cap \mathbf{R}_{\xi+1} \in \tau_{\xi+1}$, so there is an $n \in \omega$ such that $U_n \subseteq V \cap \mathbf{R}_{\xi+1} \subseteq V$, as desired. Finally, for (8_β), it suffices to notice that if $K \subseteq \mathbf{R}_\xi$ is compact in τ_ξ, where $\xi < \beta$, then it is compact in τ_β also.

Finally, suppose that β is an infinite successor ordinal $\alpha + 1$. If there is no $\mu < \alpha$ such that for some $\xi \leq \alpha$ we have $(x_\alpha, x_\xi) \in \bar{S}^\rho_\mu$ or $(x_\xi, x_\alpha) \in \bar{S}^\rho_\mu$, let τ_β be the topology with the base $\tau_\alpha \cup \{\{x_\alpha\}\}$. The conditions $(2_\beta) - (8_\beta)$ are easy to check.

Now suppose there is such a μ. Let T be the set of all ordered triples $(\gamma, \varepsilon, \mu)$ such that $\gamma, \varepsilon \leq \alpha$, $\gamma = \alpha$ or $\varepsilon = \alpha$, and $(x_\gamma, x_\varepsilon) \in \bar{S}^\rho_\mu$, where $\mu < \alpha$. Thus $0 < |T| \leq \omega$. Let $\langle (\xi_m, \eta_m, \mu_m) : m < \omega \rangle$ emumerate T, each element of T repeated infinitely many times. For each $\xi \leq \alpha$, let $\langle U^\xi_n : n < \omega \rangle$ be a decreasing sequence of open sets forming a neighborhood base for x_ξ in the usual topology ρ. Now for each $n < \omega$ choose $(p_n, q_n) \in S_{\mu_n} \cap (U_n^{\xi_n} \times U_n^{\eta_n})$. By (8_α) we find compact open (in τ_α) $K_n \subseteq U_n^\alpha$ such that $p_n \in K_n$ if $\xi_n = \alpha$ and $q_n \in K_n$ if $\eta_n = \alpha$. Let τ_β be the topology on $\mathbf{R}_\beta$ having as a base the sets in τ_α together with all sets of the form $\{x_\alpha\} \cup \bigcup_{m>n} K_m$ for $n \in \omega$. We proceed to check $(2_\beta) - (8_\beta)$. (2_β) and (3_β) are clear. For (4_β), suppose that V is open in ρ. If $x_\alpha \notin V$, then $V \cap \mathbf{R}_\beta = V \cap \mathbf{R}_\alpha$ and so $V \in \tau_\beta$. Suppose that $x_\alpha \in V$. Choose $n \in \omega$ such that $U_n^\alpha \subseteq V$. Then $\bigcup_{m>n} K_m \subseteq U_n^\alpha \subseteq V$. Hence $V \cap \mathbf{R}_\beta = (V \cap \mathbf{R}_\alpha) \cup \{x_\alpha\} \cup \bigcup_{m>n} K_m \in \tau_\beta$, proving (4_β). For (5_β), assume that $\xi, \xi' < \beta$, $\mu < \xi$ or $\mu < \xi'$, and $(x_\xi, x_{\xi'}) \in \bar{S}_\mu^\rho$. We want to show that $(x_\xi, x_{\xi'}) \in \bar{S}_\mu^{\tau_\beta}$. To this end, take a neighborhood of $(x_\xi, x_{\xi'})$; we may assume that it has the form $W \times W'$ with W and W' open in τ_β. There are three possiblities. If $\xi = \alpha$ and $\xi' < \alpha$, we proceed as follows. We may assume that W has the form $\{x_\alpha\} \cup \bigcup_{m>n} K_m$ and W' has the form $U_a^{\xi'}$ for some $n, a \in \omega$. Choose $r > n, a$ so that $(\xi_r, \eta_r, \mu_r) = (\xi, \xi', \mu)$. Then $(p_r, q_r) \in S_\mu \cap (W \times W')$, as desired. The other two cases are very similar. So (5_β) is established. Condition (6_β) is obvious, as is (7_β). For (8_β), note that a set which is compact open in τ_α remains so in τ_β. Hence it suffices to show that $\{x_\alpha\} \cup \bigcup_{m>n} K_m$ is compact for each $n \in \omega$. Suppose that $\mathcal{O}$ is an open cover of this set. Choose $V \in \mathcal{O}$ such that $x_\alpha \in V$. Then there is a $p \in \omega$ such that $\{x_\alpha\} \cup \bigcup_{m>p} K_m \subseteq V$, and without loss of generality $n < p$. Since $\mathcal{O}\backslash\{V\}$ covers $\bigcup_{n<m\leq p} K_m$, which is compact, there is a finite subset of $\mathcal{O}$ which covers the desired set $\{x_\alpha\} \cup \bigcup_{m>n} K_m$. This finishes the construction of the topologies.

For brevity, let $\tau = \tau_{\omega_1}$. To proceed further, we need the following fact about the construction:

(9) If $A \subseteq \mathbf{R} \times \mathbf{R}$, then $|\bar{A}^\rho \backslash \bar{A}^\tau| \leq \omega$.

For, let B be countable and ρ-dense in A; thus $\bar{A}^\rho = \bar{B}^\rho$. Choose $\mu < \omega_1$ so that $B = S_\mu$. By condition (5_{ω_1}) we clearly have

$$\bar{A}^\rho \backslash \bar{A}^\tau \subseteq \bar{B}^\rho \backslash \bar{B}^\tau \subseteq \{x_\xi : \xi \leq \mu\} \times \{x_\xi : \xi \leq \mu\}.$$

and (9) follows.

(10) $(\mathbf{R}, \tau) \times (\mathbf{R}, \tau)$ is hereditarily separable.

To prove (10), let X be any subspace of $(\mathbf{R}, \tau) \times (\mathbf{R}, \tau)$. Let C be a countable subset of X which is ρ-dense in X. Then $C \cup (X \backslash \bar{C}^\tau) \subseteq C \cup (\bar{C}^\rho \backslash \bar{C}^\tau)$, so $C \cup (X \backslash \bar{C}^\tau)$ is countable by (9). It is τ-dense in X, since if $U, V \in \tau$ and $(U \times V) \cap X \neq 0$, then $(U \times V) \cap X \cap (X \backslash \bar{C}^\tau) = 0$ implies that $(U \times V) \cap X \subseteq \bar{C}^\tau$ and hence $(U \times V) \cap X \cap C \neq 0$, as desired.

Now we go to the final step in this example: let Y be the one-point compactification of $(\mathbf{R}, \tau)$. Recall that Y is obtained from $\mathbf{R}$ by adding just one point, say y, where the topology on Y consists of the members of τ plus all sets of the form $\{y\} \cup U$ such that $U \subseteq \mathbf{R}$ and $\mathbf{R} \backslash U$ is compact in τ.

(11) In Y, the closed-open sets form a base for the topology.

To show this, first note that any subset of $\mathbf{R}$ which is compact in the τ sense is also compact in Y. So, it suffices to show that each "new" basic open set contains a new basic open set which is closed. So, let W be a "new" basic open set—say $W = \{y\} \cup U$, where $U \subseteq \mathbf{R}$ and $\mathbf{R} \backslash U$ is compact in τ. Since the compact open sets form a base in τ by (8_{ω_1}), it follows that $\mathbf{R} \backslash U \subseteq V$ for some $V \subseteq \mathbf{R}$ which is compact open in τ (using the compactness of $\mathbf{R} \backslash U$). Thus $\{y\} \cup \mathbf{R} \backslash V \subseteq W$ and $\{y\} \cup \mathbf{R} \backslash V$ is closed-open in Y, as desired.

(12) Y is Hausdorff.

This follows on general grounds, since $(\mathbf{R}, \tau)$ is Hausdorff and locally compact.

Thus Y is a Boolean space. It is straightforward to check that the BA of closed-open sets is uncountable (new compact-open sets were introduced at each successor step). It is clear from (10) that $Y \times Y$ is hereditarily separable. That the dual of Y has countable irredundance follows from the following result of Heindorf [89] (upon noticing that $\mathrm{s}A = \mathrm{c}_{\mathrm{H+}}A \leq \mathrm{d}_{\mathrm{H+}}A = \mathrm{hd}A$ using 1.3 and 3.6):

Theorem 6.2. *Let X be a Boolean space, and A its BA of closed-open sets. Then* $\mathrm{Irr}A \leq \mathrm{s}(X \times X)$.

Proof. Suppose that I is an infinite irredundant subset of A; we will produce an ideal independent subset of $A \times A$ of power $|I|$ (as desired—see Theorem 1.2). Namely, take the set $\{a \times -a : a \in I\}$; it is as desired, for suppose that

$$a \times -a \subseteq (b_0 \times -b_0) \cup \ldots \cup (b_{m-1} \times -b_{m-1}),$$

where $a, b_0, \ldots, b_{m-1}$ are distinct elements of I. Now a is not in $\langle \{b_i : i < m\} \rangle$, so it follows that in that subalgebra, a splits some atom; this means that there is an $\varepsilon \in {}^m 2$ such that, if we set $d = \bigcap_{i<m} b_i^{\varepsilon_i}$ then we have $d \cap a \neq 0 \neq d \cap -a$. Choosing $x \in d \cap a$ and $y \in d \cap -a$ it follows that $(x, y) \in a \times -a$ but $(x, y) \notin b_i \times -b_i$ for each $i < m$, giving the desired contradiction. □

Example 6.3. This example, which as we mentioned is from Todorčević [86], constructs a topology on a certain subset of ${}^\omega\omega$. First, some notation: If A is a set with a linear order $<$ on it, and if $k \in \omega$, then $\langle A \rangle^k$ denotes the set of all $f \in {}^k A$ such that $f_i < f_j$ for all $i < j < k$. For $f, g \in {}^\omega\omega$ define $f <^* g$ if $\exists m \forall n \geq m(fn < gn)$. The BA we want will be constructed under the assumption that there is a subset A of ${}^\omega\omega$ of power ω_1 which is unbounded under $<^*$. This is an obvious consequence of CH, but is weaker.

Without loss of generality A has order type ω_1 under $<^*$ and all members of A are strictly increasing. In fact, take the A originally given, and write $A = \{f_\alpha : \alpha < \omega_1\}$. Then one can inductively define $\bar{f}_\alpha$ for $\alpha < \omega_1$ so that $\bar{f}_\beta <^* \bar{f}_\alpha$ for $\beta < \alpha$, $f_\alpha <^* \bar{f}_\alpha$, and $\bar{f}_\alpha$ is strictly increasing. Namely, let $\bar{f}_0$ be arbitrary. If $\bar{f}_\beta$ has been constructed for all $\beta < \alpha$, let $\langle g_n : n < \omega\rangle$ enumerate $\langle \bar{f}_\beta : \beta < \alpha\rangle$. Define $\bar{f}_\alpha(n)$ to be $> \bar{f}_\alpha(m)$ for all $m < n$, also $> f_\alpha(n)$, and also $> g_m(n)$ for all $m < n$. Clearly this works. The new set $\{\bar{f}_\alpha : \alpha < \omega_1\}$ (still denoted by A below) has the desired properties.

We will apply the above notation $\langle A\rangle^k$ to A under the ordering $<^*$. Let T be an Aronszajn subtree of $\{s \in {}^{<\omega_1}\omega : s \text{ is one-one}\}$. (See Kunen [80], p. 70.) For each $\alpha < \omega_1$ let t_α be a member of T with domain α. Define $e : \langle A\rangle^2 \to \omega$ by $e(\bar{f}_\alpha, \bar{f}_\beta) = t_\beta\alpha$ for $\alpha < \beta$. Then the following conditions clearly hold:

(1) For all $b \in A$, $e_b \stackrel{\text{def}}{=} e(\cdot, b)$ is a one-one map from $A_b \stackrel{\text{def}}{=} \{a \in A : a <^* b\}$ into ω.

(2) For all $a \in A$, the set $\{e_b \restriction A_a : b \in A\}$ is countable.

For distinct $a, b \in A$ let $\Delta(a, b)$ be the least $n < \omega$ such that $an \neq bn$. And let $\Delta(a, a) = \infty$. Now we define $H : A \to \mathcal{P}A$ by

$$Hb = \{a \in A : a <^* b \text{ and } e(a, b) \leq b(\Delta(a, b))\}.$$

Note by (1) and the definition of H we have

(3) for all $l < \omega$ and $b \in A$ the set $\{a \in Hb : \Delta(a, b) = l\}$ is finite.

Next we define Cb for $b \in A$ by recursion on b: $a \in Cb$ iff $a = b$ or

(4) $\exists c \in Hb(a \in Cc \text{ and } \forall d \in Hb(d \neq a \text{ and } d \neq c \Rightarrow \Delta(a, d) < \Delta(a, c)))$.

Note that

(5) $$Hb \subseteq Cb$$

for all $b \in A$ (take $c = a$ and note that $\Delta(a, a) = \infty$).

For each $n \in \omega$ and $b \in A$ let $C_n b = \{a \in Cb : \Delta(a, b) \geq n\}$. Then

(6) $$c \in Hb \Rightarrow \exists l(C_l c \subseteq Cb).$$

In fact, $\{x \in Hb : \Delta(x, b) = \Delta(c, b)\}$ is finite by (3). Choose $l > \Delta(x, c)$ for any $x \neq c$ which is in this set, and with $l > \Delta(c, b)$. Suppose that $d \in C_l c$. We claim that $d \in Cb$, and that the element c works for this purpose in (4). Indeed, suppose that $x \in Hb$, $x \neq d$, $x \neq c$, and $\Delta(d, x) \geq \Delta(d, c)$. So x agrees with c up through $l-1$, hence $\Delta(x,b)=\Delta(c,b)$, so that x is in the indicated set, which gives a contradiction $l>\Delta(x,c)$.

$$(7)\qquad a\in Cb\Rightarrow \exists l(C_l a\subseteq Cb).$$

For, we may assume that $a\notin Hb$ by (6), and we proceed by induction on b. The conclusion is clear if $a=b$, so suppose that $a\neq b$. Choose c in accordance with (4). Then $a\neq c$ since $a\notin Hb$. By the induction hypothesis, choose l such that $C_l a\subseteq Cc$. Without loss of generality, $l>\Delta(a,c)$. We claim that $C_l a\subseteq Cb$. To prove this, let $d\in C_l a$. So, $d\in Cc$. Suppose that $x\in Hb$, $x\neq d$, $x\neq c$, and $\Delta(d,x)\geq\Delta(d,c)$. Then $x\neq a$ since $a\notin Hb$. And $\Delta(d,c)=\Delta(a,c)$ since $\Delta(d,a)>\Delta(a,c)$, so $\Delta(a,x)\geq\Delta(a,c)$, contradiction.

From (7) we immediately get

$$(8)\qquad a\in C_m b\Rightarrow \exists l(C_l a\subseteq C_m b).$$

From (8) it immediately follows that the collection of sets $\{C_m b : b\in A, m\in\omega\}$ forms a base for a topology on A. It is Hausdorff, since, given $a\neq b$, let $l=\Delta(a,b)+1$; clearly $C_l a\cap C_l b=0$. Also note that each set $Cb=C_0 b$ is open. Next,

$$(9)\qquad C_l b \text{ is closed in } Cb.$$

For, suppose that $x\in Cb\backslash C_l b$, and let $m=\Delta(x,b)+1$. Then clearly $C_m x\cap Cb\subseteq Cb\backslash C_l b$, as desired.

$$(10)\qquad Cb \text{ is compact.}$$

We prove this by induction on b. So, assume that it is true for all $c<^* b$, and suppose that $Cb\subseteq\bigcup_{x\in X}C_{m(x)}x$. Then choose $y\in X$ such that $b\in C_{m(y)}y$. There is an l such that $C_l b\subseteq C_{m(y)}y$. Now we consider two cases:

Case 1. Hb is finite. In this case, we can easily show that Cb is closed: suppose that $a\in A\backslash Cb$. Hence $a\neq b$ and

$$(*)\qquad \forall c\in Hb(a\notin Cc \text{ or } \exists d\in Hb(d\neq a \text{ and } d\neq c \text{ and } \Delta(a,d)\geq\Delta(a,c))).$$

If $c\in Hb$ and $a\notin Cc$, choose an open neighborhood U_c of a with the property that $U_c\cap Cc=0$, using the inductive hypothesis. For $c\in Hb$ and $a\in Cc$, choose $d=d(a,c)\in Hb$ such that $d\neq a$, $d\neq c$, and $\Delta(a,d)\geq\Delta(a,c)$. Let

$$V=C_{\Delta(a,b)+1}a\cap\bigcap_{c\in Hb,a\notin Cc}U_c\cap\bigcap_{c\in Hb,a\in Cc}C_{\Delta(a,d(a,c))+1}a.$$

Clearly $V\cap Cb=0$, as desired (showing that Cb is closed).

Now for each $c \in Hb$ we have that $Cc \cap Cb$ is a closed subset of Cc, and hence the inductive hypothesis finishes this case. (Here one should note that $Cb = \{b\} \cup \bigcup_{c \in Hb}(Cc \cap Cb)$.)

Case 2. Hb is infinite. Let $Y = \{c \in Hb : \Delta(c, b) < l\}$. Thus Y is finite by (3). For each $c \in Y$ let

$$D_c = \{a : \Delta(a,b) < l,\ a \in Cc, \text{ and } \forall d \in Hb(d \neq a \text{ and } d \neq c \Rightarrow \Delta(a,d) < \Delta(a,c))\}.$$

Then we claim

$$(**) \qquad c \in Y \text{ and } a \in D_c \text{ and } m = \Delta(a,c) \Rightarrow C_m c \subseteq Cb.$$

For, assume the hypotheses. If $\Delta(a,c) \leq \Delta(a,b)$, then $\Delta(d,b) < \Delta(a,b)$ for all $d \in Hb \backslash \{a,c\}$, which implies that Hb is finite, contradiction. Thus $\Delta(a,c) > \Delta(a,b)$. Thus if $u \in Cc$ and $\Delta(u,c) \geq \Delta(a,c)$, then c works for u to show that $u \in Cb$. This proves (**).

For $c \in Y$ with $D_c \neq 0$ let $m(c) = \min\{\Delta(a,c) : a \in D_c\}$. Then $Cb = C_l b \cup \bigcup_{c \in Y, D_c \neq 0} C_{m(c)}c$, and the inductive hypothesis applies to show that Cb is compact. So, we have proved (10).

From (9) and (10) we get

$$(11) \qquad C_l b \text{ is compact; so } A \text{ is locally compact.}$$

$$(12) \qquad a <^* b \Rightarrow Ca \neq Cb.$$

This is true because $b \in Cb \backslash Ca$. So there are uncountably many compact open sets.

A subset $F \subseteq \langle A \rangle^k$ is *cofinal in* A provided that for all $a \in A$ there is an $f \in F$ such that $a <^* f_i$ for all $i < k$. Next we prove

$$(13) \quad \forall \text{ finite } k \geq 1 \text{ and } \forall \text{ cofinal } F \subseteq \langle A \rangle^k \exists f, g \in F(f_i \in Hg_i \text{ for all } i < k).$$

This will take a while to prove, but it leads us close to the end of the matter. Let $D \subseteq F$ be countable dense in F in the ordinary topology (${}^\omega\omega$ has the product topology with ω having the discrete topology; ${}^k({}^\omega\omega)$ gets the product topology too). Choose $c \in A$ such that $f_i <^* c$ for all $f \in D$ and $i < k$. Now

$$\forall f \in F \exists m \in \omega \forall n \geq m \forall i < j < k(f_i n < f_j n).$$

Hence there is an uncountable $F_0 \subseteq F$ and an m_0 such that

$$(14) \qquad \forall f \in F_0 \forall n \geq m_0 \forall i < j < k(f_i n < f_j n).$$

Next, let $F_1 = \{f \in F_0 : c <^* f_0\}$. Then

$$\forall f \in F_1 \exists m > m_0 \forall n \geq m(c_n < f_0 n).$$

Hence there is an uncountable $F_2 \subseteq F_1$ and an $m > m_0$ such that

$$\forall f \in F_2 \forall n \geq m(c_n < f_0 n). \tag{15}$$

Now

$$F_2 = \bigcup \{\{f \in F_2 : \forall i < k(f_i \restriction m = s_i\} : s \in {}^k({}^m\omega)\}.$$

Hence there is an uncountable subset F_3 of F_2 and an $s \in {}^k({}^m\omega)$ such that

$$\forall f \in F_3 \forall i < k(f_i \restriction m = s_i). \tag{16}$$

Let $C = \{e_b \restriction A_c : b \in A\}$. Then

$$F_3 = \bigcup \{\{f \in F_3 : \forall i < k(e_{f_i} \restriction A_c = u^i)\} : u \in {}^kC\},$$

so there is an uncountable $F_4 \subseteq F_3$ and a $u \in {}^kC$ such that

$$\forall f \in F_4 \forall i < k(e_{f_i} \restriction A_c = u^i). \tag{17}$$

Now there is an $n \in \omega$ such that $\{f_0 n : f \in F_4\}$ is unbounded in ω, since otherwise, for each $n \in \omega$ let gn be greater than each $f_0 n$ for $f \in F_4$. It follows that g is an upper bound for A, contradiction. Take $m0$ to be the least such n. Then there is a $p \in \omega$ such that $\{f_0 \restriction m0 : f \in F_4\} \subseteq {}^{m0}p$; so there is a $t_0 \in {}^{m0}p$ and an infinite subset F_5 of F_4 such that $f_0 \restriction m0 = t_0$ for all $f \in F_5$ and $\{f_0 m0 : f \in F_5\}$ is unbounded in ω. We may assume, in fact, that F_5 has the form $\{f^{0j} : j < \omega\}$, where $f^{00}m0 < f^{01}m0 < \cdots$. Now for any $i \in \omega$, if $n = \max\{m0, m\}$, choose $f \in F_5$ such that $i < f_0 m0$; then $i < f_0 n < f_1 n$. This means that $\{f_1 n : f \in F_5\}$ is unbounded in ω; choosing the least such n and proceeding as before, and then doing the same thing for $2, \ldots, k-1$, we end up with $m0, \ldots, m(k-1)$, $t_0, \ldots, t_{k-1}$, $t_i \in {}^{mi}\omega$ for all $i < k$, $F_l = \{f^{lj} : j < \omega\}$ with $f^{l0}ml < f^{l1}ml < \cdots$ for $5 \leq l < 5+k$, and $F_4 \supseteq F_5 \supseteq \cdots \supseteq F_{4+k}$. Then the following two conditions hold:

$$\forall f \in F_{4+k} \forall i < k(t_i \subseteq f_i). \tag{18}$$

$$\forall n < \omega \exists f \in F_{4+k} \forall i < k(f_i mi > n). \tag{19}$$

For, (18) is clear. Concerning (19), let $n < \omega$. For each $i < k$, the set G_i of $f \in F_{5+i}$ such that $f_i mi \leq n$ is finite; choosing $f \in F_{4+k}$ not in any of these sets G_i gives the desired f in (19).

Now the open set in ${}^k({}^\omega\omega)$ determined by $\langle t_0, \ldots, t_k \rangle$ meets F, since F_4 is contained in it; so by the denseness of D, choose $d \in D$ in this open set: $t_i \subseteq d_i$ for all $i < k$. Then pick $\overline{m} \in \omega$ such that

$$\forall i < k \forall n \geq \overline{m} (d_i n < c_n). \tag{20}$$

By (19) there is an $f \in F_{4+k}$ such that

$$\forall i < k [f_i m i > \max(\{c_{\overline{m}}\} \cup \{u^j d_j : j < k\})].$$

Hence for all $i < k$ we have

$$e(d_i, f_i) = u^i d_i < f_i m i = f_i(\Delta(d_i, f_i)),$$

so d_i is in Hf_i for all $i < k$, as desired; we have proved (13)!

Next,

(21) For every positive integer k, the space ${}^kA$ does not have an uncountable discrete subspace.

We prove this by induction on k; suppose that it is true for all $k' < k$. Suppose that F is an uncountable discrete subspace of ${}^kA$. We may assume that there is an integer m such that $(C_m f_0 \times \cdots \times C_m f_{k-1}) \cap F = \{f\}$ for all $f \in F$. For all $f \in F$ define $\equiv_f$ on k by $i \equiv_f j$ iff $f_i = f_j$. Without loss of generality, $\equiv_f$ is the same for all $f \in F$. Hence by the induction hypothesis, $\equiv$ is the identity relation, so that each $f \in F$ is one-one. And then by a similar argument with permutations of k we may assume that $f_i <^* f_j$ whenever $f \in F$ and $i < j < k$. Next, we may assume that $\langle \text{rng} f : f \in F \rangle$ is a Δ-system, say with kernel G. For each $f \in F$, $\{i < k : f_i \in G\}$ is a finite subset of k; we may assume that this set is the same for all $f \in F$; call it Γ. Thus $f_i = g_i$ for all $f, g \in F$ and all $i \in \Gamma$. So the set $F' \stackrel{\text{def}}{=} \{f \restriction (k \backslash \Gamma) : f \in F\}$ is still uncountable and discrete. Since $\langle \text{rng} f : f \in F' \rangle$ is a system of disjoint sets, F' is cofinal in A. And

$$F' = \bigcup \{\{f \in F : \forall i < k (f_i \restriction m = s_i)\} : s \in {}^k({}^m\omega)\},$$

so we may assume that $f_i \restriction m = g_i \restriction m$ for all $f, g \in F'$ and $i < k$. Now we apply (13) to get distinct $f, g \in F'$ such that $f_i \in Hg_i$ for all $i < k$. Since $Ha \subseteq Ca$ for all a, this clearly is a contradiction. So (21) holds.

The only remaining step is to take the one-point compactification A' of A; it is clear that it satisfies the desired conditions. □

Problem 21. *Can one construct in ZFC a BA A such that* Irr$A < |A|$*?*

7. CARDINALITY

We denote $|A|$ also by Card$A$. The behaviour of this function under algebraic operations is for the most part obvious. Note, though, that questions about its behaviour under ultraproducts are the same as the well-known and difficult problems concerning the cardinality of ultraproducts in general. $\mathrm{Card}_{\mathrm{H}-}$ is a non-obvious function. Clearly $\mathrm{Card}_{\mathrm{H}-}A \leq 2^\omega$ for every infinite BA, and $\mathrm{Card}_{\mathrm{H}-}A = \omega$ for many BAs, e.g. for free BAs and interval algebras. But $\mathrm{Card}_{\mathrm{H}-}A = 2^\omega$ for A satisfying CSP. W. Just [88] has shown that it is consistent to have a BA A such that $\omega_1 \leq \mathrm{Card}_{\mathrm{H}-}A = |A| < 2^\omega$. Questions about $\mathrm{Card}_{\mathrm{H}-}$ are connected to some problems about cofinality and related cardinal functions which will not be considered here; see van Douwen [89]. The cardinal function $\mathrm{Card}_{\mathrm{h}+}$ is defined as follows:

$$\mathrm{Card}_{\mathrm{h}+}A = \sup\{|\mathrm{Clop}X| : X \subseteq \mathrm{Ult}A\}.$$

It is possible to have $\mathrm{Card}_{\mathrm{h}+}A > |\mathrm{Ult}A|$: this is true, for example, with A the finite-cofinite algebra on an infinite cardinal κ, taking X to be the set of all principal ultrafilters on A, so that X is discrete and hence $\mathrm{Clop}X = \mathcal{P}X$ and $\mathrm{Card}_{\mathrm{h}+}A = 2^\kappa$. On the other hand, $\mathrm{Card}_{\mathrm{h}-}$ coincides with $\mathrm{Card}_{\mathrm{H}-}$: obviously $\mathrm{Card}_{\mathrm{h}-}A \leq \mathrm{Card}_{\mathrm{H}-}A$, and if X is any infinite subset of Ult$A$, then the function f such that $fa = \mathcal{S}a \cap X$ is a homomorphism from A onto an algebra B such that $|B| \leq \mathrm{Clop}X$; so $\mathrm{Card}_{\mathrm{H}-}A \leq \mathrm{Card}_{h-}A$. Clearly ${}_{\mathrm{d}}\mathrm{Card}_{\mathrm{S}+}A = |A|$, and ${}_{\mathrm{d}}\mathrm{Card}_{\mathrm{S}-}A = \pi A$.

We shall now go into some detail concerning the spectrum function $\mathrm{Card}_{\mathrm{Hs}}$, which seems to be another interesting derived function associated with cardinality. First we note some more-or-less obvious facts: (1) If A is an infinite free BA, then $\mathrm{Card}_{\mathrm{Hs}}A = [\omega, |A|]$; (2) If A is infinite and complete, then $\mathrm{Card}_{\mathrm{Hs}}A = [\omega, |A|] \cap \{\kappa : \kappa^\omega = \kappa\}$ (using in an essential way the Balcar-Franěk theorem); (3) if $\omega \leq \kappa \leq |A|$, then $\mathrm{Card}_{\mathrm{Hs}}A \cap [\kappa, 2^\kappa] \neq 0$; (4) if A has a free subalgebra of power $\kappa \geq \omega$, then $\mathrm{Card}_{\mathrm{Hs}}A \cap [\kappa, \kappa^\omega] \neq 0$. Now we prove a few more involved things.

Lemma 7.1. *If κ is an infinite cardinal, L is a linear ordering, the sequence $\langle a_\alpha : \alpha < \kappa\rangle$ is strictly increasing in L, and A is the interval algebra on L, then $[\omega, \kappa] \subseteq \mathrm{Card}_{\mathrm{Hs}}A$.*

PROOF. It suffices to show that $\kappa \in \mathrm{Card}_{\mathrm{Hs}}A$. Define $x \equiv y$ iff $x, y \in L$ and $\forall\alpha < \kappa[(a_\alpha < x$ iff $a_\alpha < y)$ and $(x < a_\alpha$ iff $y < a_\alpha)]$. Then $\equiv$ is a convex equivalence relation on L with the equivalence classes of order type κ or $\kappa + 1$, and the desired homomorphism is easy to define. □

Corollary 7.2. *If κ is an infinite cardinal and A is the interval algebra on κ, then* $\mathrm{Card}_{\mathrm{Hs}}A = [\omega, \kappa]$. □

Corollary 7.3. *If κ is an infinite cardinal, L is a linear ordering of power $\geq (2^\kappa)^+$, and A is the interval algebra on L, then $[\omega, \kappa^+] \subseteq \mathrm{Card}_{\mathrm{Hs}} A$.*

PROOF. One can apply the Erdös-Rado theorem $(2^\kappa)^+ \to (\kappa^+)^2_\kappa$ to get a chain in L of order type κ^+ or $(\kappa^+)^*$. □

Theorem 7.4. *Let A be the interval algebra on* $\mathbf{R}$. *Then* $\mathrm{Card}_{\mathrm{Hs}} A = \{\omega, 2^\omega\}$.

PROOF. The inclusion $\supseteq$ is obvious. Now suppose that f is a homomorphism of A onto an uncountable BA B; we want to show that $|B| = 2^\omega$. Notice that f is determined by a convex equivalence relation E on $\mathbf{R}$, where the number of E-equivalence classes is $|B|$. Now $L' \overset{\text{def}}{=} \bigcup\{k : k$ is an E-equivalence class with $|k| > 1\}$ is Borel, so $L'' \overset{\text{def}}{=} \mathbf{R} \backslash L'$ is also. There are only countably many E-equivalence classes k such that $|k| > 1$, so clearly $|L''| = |B|$. Hence $|B| = 2^\omega$ by the Alexandroff-Hausdorff theorem (see Kuratowski [58] Theorem 3, p. 355). □

Theorem 7.5. $\mathrm{Card}_{\mathrm{Hs}}(A \times B) = \mathrm{Card}_{\mathrm{Hs}} A \cup \mathrm{Card}_{\mathrm{Hs}} B$.

PROOF. The inclusion $\supseteq$ is obvious. Now suppose that f is a homomorphism of $A \times B$ onto an infinite BA C. Let F be an ultrafilter on A and G an ultrafilter on B. Then we can define an isomorphism g of A into $A \times B$ by setting $ga = (a, a/G)$ for all $a \in A$. Similarly we get an isomorphism h of B into $A \times B$. We claim that C is generated by $(f \circ g)[A] \cup (f \circ h)[B] \cup \{f(1,0)\}$ (hence one of $(f \circ g)[A]$ or $(f \circ h)[B]$ is of power $|C|$, giving the other inclusion). To see this, let $c \in C$, and choose $(a, b) \in A \times B$ such that $f(a,b) = c$. Then $fga \cdot f(1,0) + fhb \cdot -f(1,0) = c$, as desired. □

Theorem 7.6. $\mathrm{Card}_{\mathrm{Hs}}(A \oplus B) = \mathrm{Card}_{\mathrm{Hs}} A \cup \mathrm{Card}_{\mathrm{Hs}} B$.

PROOF. The inclusion $\supseteq$ is obvious. If f is a homomorphism from $A \oplus B$ onto C, then $f[A] \cup f[B]$ generates C, so the other inclusion follows. □

Corollary 7.7. *If $\omega \leq \kappa \leq 2^\omega$, then there is a BA A such that* $\mathrm{Card}_{\mathrm{Hs}} A = [\omega, \kappa] \cup \{2^\omega\}$.

PROOF. Apply Theorem 7.5 to $A \times \mathcal{P}\omega$, where A is the free BA on κ free generators. □

Theorem 7.8 (CH). *If there is a BA A such that* $\mathrm{Card}_{\mathrm{Hs}} A = \{\omega, \omega_2\}$, *then there is an ω_1-Kurepa family.*

(For the definition of a Kurepa family and for various results concerning this notion, see Kunen [80].)

Proof. By the fact (4) above, and CH, A has no uncountable independent subset. Hence by Shelah's theorem (Part I of the Handbook, Corollary 10.9) and CH, A does not satisfy ccc. Hence A has a homomorphic image B which is a subalgebra of $\mathcal{P}\omega_1$ containing all singletons. Clearly $\mathrm{Card}_{\mathrm{Hs}}B = \{\omega, \omega_2\}$. If Γ is any countable subset of ω_1, then $b \mapsto b \cap \Gamma$ for $b \in B$ is a homomorphism, and hence $\{b \cap \Gamma : b \in B\}$ is countable. Thus B itself is an ω_1-Kurepa family. □

Two remarks concerning Theorem 7.8: If K is a Kurepa family, then the subalgebra $\langle K \rangle$ of $\mathcal{P}\omega_1$ generated by K is also a Kurepa family, since for any countable Γ the function $f : b \mapsto b \cap \Gamma$ is a homomorphism from $\langle K \rangle$ onto a subalgebra B of $\mathcal{P}\Gamma$, and B is generated by $\{fb : b \in K\}$, which is countable. The second remark is that it is consistent with CH that there is no Kurepa family, hence no BA as described in Theorem 7.8.

Problem 22. *Is it consistent with CH that there is a BA A such that* $\mathrm{Card}_{\mathrm{Hs}}A = \{\omega, \omega_2\}$?

Note that the relations $\mathrm{Card}_{\mathrm{Sr}}$ and $\mathrm{Card}_{\mathrm{Hr}}$ are trivial. In comparing cardinality with the cardinal functions so far introduced, we now note explicitly that $2^{\mathrm{d}A} \geq |A|$ for any infinite BA A. Finally, recall from Part I of the BA Handbook, Theorem 12.2, that $|A|^\omega = |A|$ for any infinite CSP algebra A, in particular for any (countably) complete infinite BA A.

8. INDEPENDENCE

There is a lot of information about independence in Part I of the Handbook. An even more extensive account is in Monk [83].

To treat the attainment problem, it is again convenient to first talk about independence in products.

Theorem 8.1. *If neither A nor B has an independent set of power $\kappa \geq \omega$ then $A \times B$ also does not.*

PROOF. Let $\langle (a_\alpha, b_\alpha) : \alpha < \kappa \rangle$ be a system of elements of $A \oplus B$; we want to show that this system is dependent. Choose a finite subset Γ of κ and $\varepsilon \in {}^\Gamma 2$ so that $\prod_{\alpha \in \Gamma} a_\alpha^{\varepsilon\alpha} = 0$, and then choose a finite subset Δ of $\kappa \backslash \Gamma$ and $\delta \in {}^\Delta 2$ so that $\prod_{\alpha \in \Delta} b_\alpha^{\delta\alpha} = 0$. Let $\Theta = \Gamma \cup \Delta$ and $\theta = \delta \cup \varepsilon$. Then $\prod_{\alpha \in \Theta}(a_\alpha, b_\alpha)^{\theta\alpha} = 0$, as desired. □

Corollary 8.2. $\mathrm{Ind}(A \times B) = \max(\mathrm{Ind}A, \mathrm{Ind}B)$ if A and B are infinite BAs. □

Corollary 8.3. *If $\langle A_i : i \in I \rangle$ is a system of BAs, κ is an infinite cardinal, and for every $i \in I$, the set A_i does not have an independent subset of power κ, then $\prod^{\mathrm{w}}_{i \in I} A_i$ also has no such subset.*

PROOF. Suppose that X is an independent subset of $\prod^{\mathrm{w}}_{i \in I} A_i$ of power κ. Fix $x \in X$. We may assume that $F \stackrel{\mathrm{def}}{=} \{i \in I : x_i \neq 0\}$ is finite. Then

$$\langle y \upharpoonright \prod_{i \in F} A_i : y \in X \backslash \{x\} \rangle$$

is a system of κ independent elements of $\prod_{i \in F} A_i$, contradicting Theorem 8.1. □

Corollary 8.4. $\mathrm{Ind}(\prod^{\mathrm{w}}_{i \in I} A_i) = \sup_{i \in I} \mathrm{Ind}A_i$. □

Corollary 8.3 enables us to take care of the attainment problem for independence. For each limit cardinal κ there is a BA A with independence κ not attained. For $\kappa = \omega$ we simply take for A any infinite superatomic BA. Now assume that κ is an uncountable limit cardinal. Let I be the set of all infinite cardinals $< \kappa$, and for each $\lambda \in I$ let B_λ be the free BA with λ free generators. Then $A \stackrel{\mathrm{def}}{=} \prod^{\mathrm{w}}_{\lambda \in I} B_\lambda$ is as desired, by Corollary 8.4.

It is perhaps surprising that the analog of Corollary 8.4 for arbitrary products is false. We give an example showing this due to T. Cramer [74].

EXAMPLE 8.5. *For each infinite cardinal κ there is a superatomic BA A such that* $\mathrm{Ind}\, {}^\omega A = \kappa$. To see how surprising this is, recall that a superatomic

BA has no infinite independent subset. To construct A, we first construct, for each positive integer n, a certain algebra B_n. Namely, B_n is the subalgebra of $\mathcal{P}([\kappa]^n)$ generated by $\{a^n_\alpha : \alpha < \kappa\}$, where for each $\alpha < \kappa$ we set

$$a^n_\alpha = \{X \in [\kappa]^n : \alpha \in X\}.$$

Also recall that for any set Y and cardinal λ, $[Y]^\lambda = \{Z \subseteq Y : |Z| = \lambda\}$. To derive important properties of B_n we notice the following:

(1) If $m \in \omega$, $\langle \alpha(i) : i < m\rangle$ is a sequence of distinct ordinals $< \kappa$, and $\varepsilon \in {}^m2$, then $\prod_{i<m}(a^n_{\alpha(i)})^{\varepsilon i} = \{X \subseteq \kappa : |X| = n,\ \alpha(i) \in X$ whenever $\varepsilon i = 1$, and $\alpha(i) \notin X$ whenever $\varepsilon i = 0\}$.

From this the following fact is clear:

(2) Under the hypotheses of (1), let $b = \prod_{i<m}(a^n_{\alpha(i)})^{\varepsilon i}$ and let $p = |\{i < m : \varepsilon i = 1\}|$; then $b \neq 0$ iff $p \leq n$; for $p < n$, b is the sum of κ atoms, and for $p = n$, b is an atom. Thus B_n is atomic.

(3) If n is a positive integer, let I be the ideal of B_{n+1} generated by its atoms. Then $B_{n+1}/I \cong B_n$.

To prove (3), an isomorphism f is determined by:

$$f[a^{n+1}_\alpha] = a^n_\alpha \text{ for all } \alpha < \kappa.$$

To see this, it suffices by Sikorski's extension criterion to note that if $\alpha(0), \ldots, \alpha(m)$ is a sequence of distinct ordinals $< \kappa$ and $\varepsilon(0), \ldots, \varepsilon(m) \in 2$, then

$$\begin{aligned}
[a^{n+1}_{\alpha(0)}]^{\varepsilon(0)} \cdot \ldots \cdot [a^{n+1}_{\alpha(m)}]^{\varepsilon(m)} = 0 &\text{ iff } (a^{n+1}_{\alpha(0)})^{\varepsilon(0)} \cap \ldots \cap (a^{n+1}_{\alpha(m)})^{\varepsilon(m)} \in I \\
&\text{ iff } |\{i \leq m : \varepsilon i = 1\}| \geq n+1 \\
&\text{ iff } (a^n_{\alpha(0)})^{\varepsilon(0)} \cap \ldots \cap (a^n_{\alpha(m)})^{\varepsilon(m)} = 0,
\end{aligned}$$

as desired.

Note that B_1 is the BA of finite and cofinite subsets of κ. Hence from (3) and Proposition 17.8 of Part I of the BA handbook it follows that each algebra is superatomic. Now we let $A = \prod^{\mathrm{w}}_{0<n<\omega} B_n$; by Example 17.18 of Part I of the BA handbook, A is also superatomic. Finally, we define $f \in {}^\kappa({}^\omega A)$ by the following rule, where $\alpha < \kappa$, $n < \omega$, and $0 < i < \omega$:

$$(f_\alpha n)i = \begin{cases} 0, & \text{if } n \neq i, \\ a^n_\alpha, & \text{if } n = i. \end{cases}$$

Clearly f is an independent system in ${}^\omega A$.

Although this example takes care of the most obvious question about independence in products, there is still an important problem left. It arises upon asking whether Example 8.5 can be done with an interval algebra (they always have independence ω too, just like superatomic algebras, although independence is attained for some interval algebras). The answer is no, but the situation is a little complicated. First observe that if $\langle A_i : i \in I\rangle$ is a system of non-trivial BAs and I is infinite, then $\mathrm{Ind}(\prod_{i\in I} A_i) \geq 2^{|I|}$; this follows from the fact that $\mathcal{P}I$, which has independence $2^{|I|}$, can be embedded in $\prod_{i\in I} A_i$. This leads to the following problem:

Problem 23. *If A_i is a non-trivial interval algebra for each $i \in I$, where I is infinite, is $\mathrm{Ind}(\prod_{i\in I} A_i) = 2^{|I|}$? Equivalently, is it true that for every infinite cardinal κ there is no linear order L and sequence $\langle x_\alpha : \alpha < (2^\kappa)^+\rangle$ with the following properties?*

(1) For all $\alpha < (2^\kappa)^+$ and $\beta < \kappa$, $x_\alpha\beta$ is a finite collection of half-open intervals of L.

(2) For all finite disjoint $\Gamma, \Delta \subseteq (2^\kappa)^+$ there is a $\beta < \kappa$ such that

$$\bigcap_{\alpha\in\Gamma}(\bigcup x_\alpha\beta) \cap \bigcap_{\alpha\in\Delta}(L\backslash\bigcup x_\alpha\beta) \neq 0.$$

There are two results concerning this problem; Shelah, Soukup [89], and the following theorem of Tim Carlson.

Theorem 8.6. *Suppose that A_i is a non-trivial interval algebra for each $i \in I$, where I is infinite, $|I| = \kappa$. Then $\mathrm{Ind}(\prod_{i\in I} A_i) \leq 2^{2^\kappa}$.*

PROOF. First we prove a general fact about dependence in interval algebras. Let L be a linearly ordered set and B the interval algebra on L. For each $x \in B$ write

$$x = [a_0^x, a_1^x) \cup \ldots \cup [a^x_{2(n(x))-2}, a^x_{2(n(x))-1}).$$

where $-\infty \leq a_0^x < a_1^x < \ldots < a^x_{2(n(x))-1} \leq +\infty$. We call $n(x)$ the *length* of x. If x and y have the same length then they are said to have *relative position* ϕ, where $\phi \in {}^{2n\times 2n}3$ is defined as follows: for any $i, j < 2n$, $\phi(i,j)$ is 0,1, or 2 according as $a_i^y < a_j^x$, $a_i^y = a_j^x$, or $a_i^y > a_j^x$. Now we want to prove the following general dependence result concerning this notion:

(1) If $x(0), x(1), x(2), x(3) \in B$ have the same length and there is a ϕ such that for all i, j with $0 \leq i < j \leq 3$, $x(i)$ and $x(j)$ have relative position ϕ, then $x(0)\cdot -x(1)\cdot x(2)\cdot -x(3) = 0$.

To prove (1) we introduce some terminology: for any $y \in L$, let $a_{-1}^y = -\infty$ if $a_0^y \neq -\infty$, and $a_{2n}^y = \infty$ if $a_{2n-1}^y \neq \infty$. Now it suffices to take arbitrary i, j, k, l with $i, k < n$ and $j, l \in n \cup \{-1\}$ and show that

(2) $[a^{x(0)}_{2i}, a^{x(0)}_{2i+1}) \cap [a^{x(1)}_{2j+1}, a^{x(1)}_{2j+2}) \cap [a^{x(2)}_{2k}, a^{x(2)}_{2k+1}) \cap [a^{x(3)}_{2l+1}, a^{x(3)}_{2l+2}) = 0.$

Suppose that (2) fails to hold. Then the elements $a^{x(0)}_{2i}, a^{x(1)}_{2j+1}, a^{x(2)}_{2k}, a^{x(3)}_{2l+1}$ are all less than each of the elements $a^{x(0)}_{2i+1}, a^{x(1)}_{2j+2}, a^{x(2)}_{2k+1}, a^{x(3)}_{2l+2}$. From the fact that relative positions are the same, this implies the following additional facts (and others which we don't need): $a^{x(2)}_{2l+1} \leq a^{x(1)}_{2i+1}$; $a^{x(2)}_{2l+1} \leq a^{x(0)}_{2j+2}$; $a^{x(3)}_{2k} \leq a^{x(1)}_{2i+1}$; $a^{x(3)}_{2k} \leq a^{x(0)}_{2j+2}$. Now we get contradictions in each of the following four possible cases:

Case 1. $i \leq j$ and $k \leq l$. A contradiction follows from

$$a^{x(2)}_{2k+1} \leq a^{x(2)}_{2l+1} \leq a^{x(1)}_{2i+1} \leq a^{x(1)}_{2j+1}.$$

Case 2. $j < i$ and $k \leq l$. A contradiction follows from

$$a^{x(2)}_{2k+1} \leq a^{x(2)}_{2l+1} \leq a^{x(0)}_{2j+2} \leq a^{x(0)}_{2i}.$$

Case 3. $i \leq j$ and $l < k$. A contradiction follows from

$$a^{x(3)}_{2l+2} \leq a^{x(3)}_{2k} \leq a^{x(1)}_{2i+1} \leq a^{x(1)}_{2j+1}.$$

Case 4. $j < i$ and $l < k$. A contradiction follows from

$$a^{x(3)}_{2l+2} \leq a^{x(3)}_{2k} \leq a^{x(0)}_{2j+2} \leq a^{x(0)}_{2i}.$$

This proves the above general fact (1) about interval algebras. To prove the theorem, suppose without loss of generality that actually $I = \kappa$, and assume that X is an independent subset of $\prod_{\alpha<\kappa} A_\alpha$ of power $(2^{2^\kappa})^+$. We may assume that there is a function $\psi \in {}^\kappa\omega$ such that for each $x \in X$ and $\alpha < \kappa$, x_α has length $\psi\alpha$. Let $\prec$ be any well ordering of X. Then $[X]^2$ can be partitioned into 2^κ classes indexed by $\prod_{\alpha<\kappa}({}^{2(\psi\alpha)\times 2(\psi\alpha)}3)$ by putting a pair $\{x, y\}$, where $x \prec y$, into Y_χ ($\chi \in \prod_{\alpha<\kappa}({}^{2(\psi\alpha)\times 2(\psi\alpha)}3)$ provided that x_α and y_α have relative position $\chi\alpha$ for all $\alpha < \kappa$. Then by the Erdös-Rado theorem $(2^\lambda)^+ \to (\lambda^+)^2_\lambda$, where $\lambda = 2^\kappa$, there exist a $\chi \in \prod_{\alpha<\kappa}({}^{2(\psi\alpha)\times 2(\psi\alpha)}3)$ and a subset Y of X of power $(2^\kappa)^+$ such that if $x(0), x(1), x(2), x(3) \in Y$ with $x(0) \prec x(1) \prec x(2) \prec x(3)$, then for any $\alpha < \kappa$ and $0 \leq i < j \leq 3$, the elements $x(i)_\alpha$ and $x(j)_\alpha$ have relative position $\chi\alpha$. But then the above fact (1) yields that $x(0) \cdot -x(1) \cdot x(2) \cdot -x(3) = 0$, contradiction. □

We turn to independence in ultraproducts. As in the case of cellularity, it is easy to see that if F is a countably complete ultrafilter on an index set I and each A_i has countable independence, then so does $\prod_{i\in I} A_i/F$. Namely, suppose that $\langle f_\alpha/F : \alpha < \omega_1\rangle$ is a system of independent elements of $\prod_{i\in I} A_i/F$. Now for every $i \in I$ there exist finite disjoint subsets $M(i), N(i)$ of ω_1 such that

$$\prod_{\alpha\in M(i)} f_\alpha i \cdot \prod_{\alpha\in N(i)} -f_\alpha i = 0.$$

Hence

$$I = \bigcup_{M,N} \{i \in I : M = M(i) \text{ and } N = N(i)\},$$

with M and N ranging over finite subsets of ω_1, so, since F is ω_2-complete, there exist finite disjoint $M, N \subseteq \omega_1$ such that $\{i \in I : M = M(i) \text{ and } N = N(i)\} \in F$. But then $\prod_{\alpha\in M} f_\alpha/F \cdot \prod_{\alpha\in N} -f_\alpha/F = 0$, contradiction.

Further, if F is countably incomplete and each algebra A_i is infinite, then $\prod_{i\in I} A_i/I$ is ω_1-saturated, hence is CSP, from which it follows that $\prod_{i\in I} A_i/I$ has independence $\geq 2^\omega$. (See Part I of the BA handbook, Theorem 13.20.) Like with cellularity, if I is infinite, F is a $|I|$-regular ultrafilter on I, and A_i is an infinite BA for each $i \in I$, then $\text{Ind}(\prod_{i\in I} A_i/I) \geq 2^{|I|}$. The proof is similar to that for cellularity: let E be a subset of F such that $|E| = |I|$ and each $i \in I$ belongs to only finitely many members of E; let Gi be the set of all $e \in E$ such that $i \in e$. With each $g \in {}^E2$ we associate $g' \in \prod_{i\in I} A_i$ as follows. Let $\langle x_h : h \in {}^{Gi}2\rangle$ be a system of independent elements of A_i. Then for any $i \in I$ we set $g'i = x_{g\restriction Gi}$. We claim that $\langle [g'] : g \in {}^E2\rangle$ is an independent system of elements of $\prod_{i\in I} A_i/I$. To see this, let $[(g0)'], \ldots, [(g(m-1))']$ be distinct elements of $\prod_{i\in I} A_i/I$ and let $\varepsilon \in {}^m2$. Let H be a finite subset of E such that $(g0) \restriction H, \ldots, (g(m-1)) \restriction H$ are all distinct. Let $i \in \bigcap H$ be arbitrary. Now $H \subseteq Gi$, so $(g0) \restriction Gi, \ldots, (g(m-1)) \restriction Gi$ are all distinct. Hence

$$((g0)'i)^{\varepsilon 0} \cdot \ldots \cdot ((g(m-1))'i)^{\varepsilon(m-1)} = x^{\varepsilon 0}_{(g0)\restriction Gi} \cdot \ldots \cdot x^{\varepsilon(m-1)}_{(g(m-1))\restriction Gi} \neq 0,$$

as desired.

Independence in free products is treated in Part I of the BA handbook: $\text{Ind}(A \oplus B) = \max(\text{Ind}A, \text{Ind}B)$, while if I is infinite and $|A_i| \geq 4$ for each $i \in I$, then $\text{Ind}(\oplus_{i\in I} A_i) = \max(|I|, \sup_{i\in I}\text{Ind}A_i)$; see Part I, Theorem 11.15. Under subalgebra and homomorphic image formation, the behaviour of independence is basically simple: if A is a subalgebra or homomorphic image of B, then $\text{Ind}A \leq \text{Ind}B$, and the difference can be arbitrarily large. Finally, independence is an ordinary sup-function, and so its behaviour with respect to unions of well-ordered chains is given by Theorem 1.1.

We turn to the functions derived from independence. $\text{Ind}_{\text{H}+}$, $\text{Ind}_{\text{S}+}$, and ${}_d\text{Ind}_{\text{S}+}$ all coincide with Ind itself. $\text{Ind}_{\text{H}-}$ appears to be a new function. In unpublished work, S. Koppelberg has constructed, using $\diamondsuit$, a BA A such that $\text{Ind}_{\text{H}-}A = \omega$ and $\text{Card}_{\text{H}-}A = \omega_1$.

Problem 24. *Can one construct in ZFC a BA A with the property that* $\mathrm{Ind}_{\mathrm{H}-}A < \mathrm{Card}_{\mathrm{H}-}A$?

Clearly $\mathrm{Ind}_{\mathrm{S}-}A = \omega$ for any infinite BA A. We define

$$\mathrm{Ind}_{\mathrm{h}+}A = \sup\{|X| : X \subseteq \mathrm{Clop}Y,\ X \text{ is independent},\ Y \subseteq \mathrm{Ult}A\}.$$

Then it is possible to have A superatomic, hence with $\mathrm{Ind}A = \omega$, while $\mathrm{Ind}_{\mathrm{h}+}A > |\mathrm{UltA}|$; see the argument for Card. In fact, maybe it is always true that $\mathrm{Ind}_{\mathrm{h}+}A = \mathrm{Card}_{\mathrm{h}+}A$:

Problem 25. *Is* $\mathrm{Ind}_{\mathrm{h}+}A = \mathrm{Card}_{\mathrm{h}+}A$ *for every infinite BA A?*

$\mathrm{Ind}_{\mathrm{h}-}$ is defined analogously. Again we do not know anything about this cardinal function:

Problem 26. *Is* $\mathrm{Ind}_{\mathrm{h}-}A = \mathrm{Ind}_{\mathrm{H}-}A$ *for every infinite BA A?*

The function ${}_{\mathrm{d}}\mathrm{Ind}_{\mathrm{S}-}$ appears to be interesting; if A is $\mathcal{P}X$ for some infinite X, then we have ${}_{\mathrm{d}}\mathrm{Ind}_{\mathrm{S}-}A = \omega$ since the BA of finite and cofinite subsets of X is dense A. On the other hand, if A is an infinite free BA of regular cardinality, then ${}_{\mathrm{d}}\mathrm{Ind}_{\mathrm{S}-}A = |A|$; see Part I of the BA handbook, Theorem 9.16.

Concerning the spectrum function $\mathrm{Ind}_{\mathrm{Hs}}$, note that if $\mathrm{Ind}_{\mathrm{H}-}A \leq \mu \leq \mathrm{Ind}A$, then A has a homomorphic image B such that $\mu \leq \mathrm{Ind}B \leq \mu^{\omega}$. Moreover, if A has CSP, then this cannot be improved:

$$\mathrm{Ind}_{\mathrm{Hs}}A = \{\lambda : 2^{\omega} \leq \lambda \leq \mathrm{Ind}A,\ \lambda^{\omega} = \lambda\}.$$

(These remarks are due to S. Koppelberg.)

The spectrum function $\mathrm{Ind}_{\mathrm{Ss}}$ is trivial: $\mathrm{Ind}_{\mathrm{Ss}}A = [\omega, \mathrm{Ind}A]$ for every infinite BA.

Caliber has been implicitly studied quite a bit for independence. We formulate one vague question in this regard. A BA A has *free caliber* κ if $\forall X \in [A]^{\kappa} \exists Y \in [X]^{\kappa}(Y \text{ is independent}\}$; $\mathrm{Freecal}A$ is the set of all $\kappa \leq |A|$ such that A has free caliber κ. It is open to characterize in purely cardinal number terms the sets Γ of cardinals such that there is a BA A such that $\mathrm{Freecal}A = \Gamma$.

The comparison of independence with the cardinal functions already introduced is simple: $\mathrm{Ind}A \leq \mathrm{Irr}A$ for every infinite BA A, and the difference can be arbitrarily large, for example in an interval algebra; it is possible to have $\mathrm{Ind}A$ bigger than πA, for example in $\mathcal{P}\kappa$. $\mathrm{Depth}A$ can be much larger than $\mathrm{Ind}A$, for example in the interval algebra on κ. Note that there are some close relationships between independence and cellularity, though. For example, $(2^{\mathrm{c}A})^+ \leq \mathrm{Ind}A$ by Corollary 10.9 of Part I of the BA handbook.

The relations Ind_{Sr} and Ind_{Hr} have not been investigated.

We close this section with some comments on independence for special kinds of BAs. By the Balcar-Franěk theorem, $\text{Ind}A = |A|$ for infinite and complete. For CSP algebras in general, all one can say is that $\text{Ind}A = (\text{Ind}A)^{\omega}$; see Part I of the BA handbook, Theorem 13.20. Finally, recall the important fact that interval algebras and superatomic algebras have countable independence.

9. π-CHARACTER

First of all, note that if F is a non-principal ultrafilter on a BA A, then $\pi\chi F \geq \omega$. To see this, suppose that X is a finite set of non-zero elements of A which is dense in F Choose $y \in F$ such that $y < \prod(X \cap F)$. Then choose $x \in X$ such that $x \leq y \cdot \prod\{z \in F : -z \in X\}$. This clearly gives a contradiction, whether $x \in F$ or not.

We begin the actual discussion with products. Clearly $\pi\chi(A \times B) = \max(\pi\chi A, \pi\chi B)$ for any infinite BAs A and B. More generally, we have:

Theorem 9.1. *For any system $\langle A_i : i \in I\rangle$ of infinite BAs we have $\pi\chi(\prod^{\mathrm{w}}_{i\in I} A_i) = \sup_{i\in I}\pi\chi A_i$.*

PROOF. We may assume that I is infinite. Since $\mathrm{Ult}(\prod^{\mathrm{w}}_{i\in I} A_i)$ is the one-point compactification of the disjoint union of all of the spaces $\mathrm{Ult}A_i$, it suffices to prove the following:

(1) Let F be the ultrafilter on $\prod^{\mathrm{w}}_{i\in I} A_i$ consisting of all $x \in \prod^{\mathrm{w}}_{i\in I} A_i$ such that $\{i \in I : x_i \neq 1\}$ is finite. Then $\pi\chi F = \omega$.

To prove (1), let J be any denumerable subset of I. For each $j \in J$ we define an element x^j of $\prod^{\mathrm{w}}_{i\in I} A_i$ by setting, for each $i \in I$,

$$x^j_i = \begin{Bmatrix} 0 \text{ if } j \neq i, \\ 1 \text{ if } j = i. \end{Bmatrix}$$

We claim that $\{x^j : j \in J\}$ is dense in F. To see this, take any $y \in F$. Then there is a $j \in J$ such that $y_j = 1$. So $x^j \leq y$, as desired. □

Note that the proof of Theorem 9.1 shows that π-character is attained in $\prod^{\mathrm{w}}_{i\in I} A_i$ iff there is an $i \in I$ such that $\pi\chi \prod_{i\in I} A_i = \pi\chi A_i$ and $\pi\chi A_i$ is attained. Using this remark, we can describe the attainment property of π-character: for each uncountable limit cardinal κ there is a BA A with π-character κ not attained: we take the weak product of free algebras of the obvious sizes. On the other hand, if $\pi\chi A = \omega$, then it is attained, since any non-principal ultrafilter has infinite π-character by our initial remark.

Turning to arbitrary products, we have:

Theorem 9.2. *If $\langle A_i : i \in I\rangle$ is a system of non-trivial BAs with $\prod_{i\in I} A_i$ infinite and $|I|$ regular, then $\pi\chi(\prod_{i\in I} A_i) \geq \max(|I|, \sup_{i\in I}\pi\chi A_i)$.*

PROOF. If $i \in I$ and G is an ultrafilter on A_i, then the following set $F \stackrel{\text{def}}{=} \{y \in \prod_{i\in I} A_i : y_i \in G\}$ is an ultrafilter on $\prod_{i\in I} A_i$, and a subset of $\prod_{i\in I} A_i$ dense in F clearly gives rise to a subset of A_i with no more elements which is dense in G. Hence $\pi\chi A_i \leq \pi\chi(\prod_{i\in I} A_i)$.

Next, assume that I is infinite; we show that $|I| \leq \pi\chi(\prod_{i\in I} A_i)$. For each subset J of I let x_J be the characteristic function of J, considered as a member of $\prod_{i\in I} A_i$. Let F be any ultrafilter on $\prod_{i\in I} A_i$ containing all elements $x_{I\setminus J}$ such that $|J| < |I|$. Then, we claim, $\pi\chi F \geq |I|$. In fact, suppose that $X \subseteq A^+$, X is dense in F, and $|X| < |I|$. For each $y \in X$ choose $i(y) \in I$ such that $y_{i(y)} \neq 0$. Let $J = \{i(y) : y \in X\}$. Then the element $x_{I\setminus J}$ of F is not $\geq$ any element of X, contradiction. □

Actually, $\pi\chi$ can jump tremendously in a product. To show this, we use the construction in Example 8.5. Recall that the algebra A there has power κ, where κ is any cardinal given in advance. We claim that $\pi\chi A = \omega$ and $\pi\chi({}^{\omega}A) = \kappa$. To see the first part of the claim, by Theorem 9.1 it suffices to show that $\pi\chi B_n = \omega$ for each positive integer n. Let F be any ultrafilter on B_n; we may assume that F is non-principal, and hence does not contain an atom. Let $M = \{\alpha < \kappa : a^n_\alpha \in F\}$. Then by the description given in Example 8.5 we must have $|M| < n$. Let N be a denumerable set of members of $[\kappa]^n$ such that $b \cap c = M$ for any two distinct members $b, c \in N$. We claim that $\{\{b\} : b \in N\}$ is a subset of B_n which is dense in F (as desired). In fact, if $b \in N$, then $\prod_{\alpha\in b} a^n_\alpha = \{b\}$ by (1) of Example 8.5, so $N \subseteq B_n$. For the denseness, suppose that $x \in F$; without loss of generality x has the form $\prod_{\alpha\in M} a^n_\alpha \cdot \prod_{\beta\in J} -a^n_\beta$ for some finite subset J of κ disjoint from M. Choosing $b \in N$ such that $b \cap J = 0$ we clearly have $b \in x$ and so $\{b\} \subseteq x$.

Now we show that $\pi\chi({}^{\omega}A) = \kappa$. With the function f as defined in Example 8.5, let F be any ultrafilter on ${}^{\omega}A$ such that $f_\alpha \in F$ for all $\alpha < \kappa$. Suppose that $X \subseteq {}^{\omega}A$ is dense in F and $|X| < \kappa$. Then there is an $x \in X$ and an infinite $M \subseteq \kappa$ such that $x \leq f_\alpha$ for all $\alpha \in M$. Choose $n \in \omega$ such that $x_n \neq 0$. Now $x_n \leq f_\alpha n$, so $x_n i = 0$ for $n \neq i$, while $x_n n \leq a^n_\alpha$ for all $\alpha \in M$, and so $\alpha \in Y$ for each $Y \in x_n n$ and each $\alpha \in M$, contradiction.

The possibility of doing the above with interval algebras, which naturally arose in section 8, is not so interesting here, since interval algebras can have high π-character (see the end of this section).

π-character has not been studied for ultraproducts.

Concerning subalgebras, it is possible to have A a subalgebra of B while $\pi\chi A > \pi\chi B$; this is true, for example, if $B = \mathcal{P}\omega$ and A is a free subalgebra of size 2^ω; then $\pi\chi B = \omega$ and $\pi\chi A = 2^\omega$. Similarly for A a homomorphic image of B: take B to be the subalgebra of $\mathcal{P}\omega$ generated by $\{\{i\} : i < \omega\} \cup M$, where M a collection of independent elements of $\mathcal{P}\omega$ of size 2^ω, and let $A = B/\text{fin}$; then A is isomorphic to the free BA of power 2^ω.

Next we describe π-character for free products:

Theorem 9.3. *If $\langle A_i : i \in I\rangle$ is a system of BAs each with at least 4 elements, then $\pi\chi \bigoplus_{i\in I} A_i = \max(|I|, \sup_{i\in I}\pi\chi A_i)$.*

PROOF. For brevity let $B = \oplus_{i\in I}A_i$. First take any $i \in I$; we show that $\pi\chi A_i \leq \pi\chi B$. Let F be any ultrafilter on A_i, and extend F to an ultrafilter G on B. Suppose $X \subseteq B$ is dense in G. We may assume that each $x \in X$ has the form

(1) $x = \prod_{j\in Mx} y_j^x$

for some finite subset Mx of I, where $y_j^x \in A_j$ for every $j \in Mx$. Now define $Y = \{y_i^x : x \in X, i \in Mx\}$. Then clearly Y is dense in F and $|Y| \leq |X|$. This proves that $\pi\chi A_i \leq \pi\chi B$.

Next, we show that $|I| \leq \pi\chi B$, where we assume that I is infinite. For each $i \in I$ choose $a_i \in A_i$ such that $0 < a_i < 1$. Let F be an ultrafilter on B such that $a_i \in F$ for each $i \in I$; clearly such an ultrafilter exists. Suppose that $X \subseteq B$ is dense in F; we may assume that each $x \in X$ has the form (1) indicated above. Clearly then, by the free product property, we must have $|X| \geq |I|$.

Now let F be an ultrafilter on B. Then for each $i \in I$, $F \cap A_i$ is an ultrafilter on A_i, and so there is an $X_i \subseteq A_i$ of cardinality $\leq \pi\chi A_i$ which is dense in $F \cap A_i$. Let

$$Y = \{y : \text{there is a finite } J \subseteq I \text{ and a } b \text{ in } \prod_{j\in J} X_j \text{ such that } y = \prod_{j\in J} b_j\}.$$

Clearly $|Y| \leq \max(|I|, \sup_{i\in I} A_i)$ and Y is dense in F, as desired. □

Concerning the behaviour of $\pi\chi$ under unions, we have the following analog of Theorem 4.1; it is proved in essentially the same way.

Theorem 9.4. *Suppose that $\langle A_\alpha : \alpha < \kappa\rangle$ is a strictly increasing sequence of BA's, with union B, where κ is regular. Let $\lambda = \sup_{\alpha<\kappa}\pi\chi A_\alpha$. Then $\kappa \leq 2^\lambda$, $\pi\chi B \leq \max(\kappa, \lambda)$, and $\pi\chi B \leq \lambda^+$.* □

Concerning the derived functions of π-character, the first result is that $\mathrm{t}A = \pi\chi_{\mathrm{H}+}A = \pi\chi_{\mathrm{h}+}A$, where $\pi\chi_{\mathrm{h}+}A = \sup\{\pi\chi(F, Y) : F \in Y, Y \subseteq \mathrm{Ult}A\}$, and for any point x of any space X, $\pi\chi(x, X)$ is defined to be $\min\{|M| : M$ is a collection of non-empty open subsets of X and for every neighborhood U of x there is a $V \in M$ such that $V \subseteq U\}$. Such a set M is called a *local π-base* for x.

Theorem 9.5. *For any infinite BA A we have $\mathrm{t}A = \pi\chi_{\mathrm{H}+}A = \pi\chi_{\mathrm{h}+}A$.*

PROOF. First we show $\mathrm{t}A \leq \pi\chi_{\mathrm{h}+}A$. For brevity let $\kappa = \pi\chi_{\mathrm{h}+}A$. Let F be an ultrafilter on A, and suppose that $Y \subseteq \mathrm{Ult}A$ and $F \subseteq \bigcup Y$; we want to find a subset Z of Y of size $\leq \kappa$ such that $F \subseteq \bigcup Z$. We may assume that $F \notin Y$. By the definition of $\pi\chi_{\mathrm{h}+}A$, let M be a local π-base for F in $Y \cup \{F\}$ with $|M| \leq \pi\chi_{\mathrm{h}+}A$. The assumption that $F \notin Y$ implies that F is not isolated in $Y \cup \{F\}$, and hence that $V \cap Y \neq 0$ for every $V \in M$. Taking a point from each such intersection, we get a subset Z of Y of power $\leq \kappa$ such that $V \cap Z \neq 0$ for every $V \in M$. Then clearly $F \subseteq \bigcup Z$, as desired.

Next we show that $\pi\chi_{\mathrm{h+}}A \leq \pi\chi_{\mathrm{H+}}A$. Given $Y \subseteq \mathrm{Ult}A$, let $\overline{Y}$ be the closure of Y, and recall from the duality theory that $\overline{Y}$ corresponds to a homomorphic image of A. So, we just need to show that $\pi\chi Y \leq \pi\chi\overline{Y}$. Let $y \in Y$, and let M be a local π-base for y in $\overline{Y}$. Then $\{U \cap Y : U \in M\}$ is clearly a local π-base for y in Y. So, $\pi\chi Y \leq \pi\chi\overline{Y}$ follows.

Finally, we show that $\pi\chi_{\mathrm{H+}}A \leq \mathrm{t}A$. Note that if Y is a closed subspace of X and $\langle x_\xi : \xi < \alpha\rangle$ is a free sequence in Y, then it is a free sequence in X also. Hence it suffices to show that if κ is any cardinal less than $\pi\chi A$ then there is a free sequence of length κ^+ in A, using Theorem 2.10. Let $F \in \mathrm{Ult}A$ with $\pi\chi F \geq \kappa^+$. Thus we have:

(1) For every subset B of A^+ of power $\leq \kappa$ there is an $a \in F$ such that $b \cdot -a \neq 0$ for every $b \in B$.

We construct a sequence $\langle a_\alpha : \alpha < \kappa^+\rangle$ by induction. Choose a_0 arbitrary $\in F$. Now suppose that a_β has been defined for all $\beta < \alpha$, where $0 < \alpha < \kappa^+$. Let G_α be the set of all non-zero products $\prod_{\beta\in M} a_\beta \cdot \prod_{\beta\in N} -a_\beta$ such that M and N are finite disjoint subsets of α such that $M < N$ (meaning that $\forall\kappa \in M\forall\lambda \in N(\kappa < \lambda)$). By (1), choose $a_\alpha \in F$ such that $b \cdot -a_\alpha \neq 0$ for all $b \in G_\alpha$.

Next for all $\alpha < \kappa^+$ let

$$H_\alpha = \{a_\beta : \beta \leq \alpha\} \cup \{-a_\beta : \alpha < \beta < \kappa^+\}.$$

By the definition of the sequence $\langle a_\alpha : \alpha < \kappa^+\rangle$, H_α has the finite intersection property. Extend H_α to an ultrafilter K_α. Then $\langle K_\alpha : \alpha < \kappa^+\rangle$ is a free sequence, since for any $\alpha < \kappa^+$ we have $\{K_\beta : \beta < \alpha\} \subseteq \mathcal{S}(-a_\alpha)$ and on the other hand $\{K_\beta : \alpha \leq \beta\} \subseteq \mathcal{S}(a_\alpha)$, as desired. □

Note from the proof of Theorem 9.5 that one of $\pi\chi_{\mathrm{h+}}$ and $\pi\chi_{\mathrm{H+}}$ is attained iff the other is; and if $\pi\chi_{\mathrm{h+}}$ is attained, then so is t, in the free sequence sense.

It is possible to have $\pi\chi_{\mathrm{S+}}A > \pi\chi A$; this is true, for example, for $A = \mathcal{P}\omega$, using the fact that $\mathcal{P}\omega$ has a free subalgebra of size 2^ω.

Clearly $\pi\chi_{\mathrm{S-}}A = \pi\chi_{\mathrm{H-}}A = \omega$. On the other hand, $\pi\chi_{\mathrm{h-}}A = 1$ for any infinite BA A, since $\mathrm{Ult}A$ has a denumerable discrete subspace. It is easily checked that ${}_{\mathrm{d}}\pi\chi_{\mathrm{S-}}A = {}_{\mathrm{d}}\pi\chi_{\mathrm{S+}}A = \pi\chi A$ for any infinite BA A.

Recall from the introduction that for a cardinal function such as $\pi\chi$ we can define an associated function $\pi\chi_{\mathrm{inf}}$ as follows: $\pi\chi_{\mathrm{inf}}A = \inf\{\pi\chi F : F$ is an ultrafilter on $A\}$. And recall from Part I Theorem 10.16 the useful result of Shapirovskiĭthat $\mathrm{Ind}A = (\pi\chi_{\mathrm{inf}})_{\mathrm{H+}}A = \sup\{\pi\chi_{\mathrm{inf}}B : B$ is a homomorphic image of $A\}$, for A not superatomic.

The functions $\pi\chi_{\mathrm{Hs}}$ and $\pi\chi_{\mathrm{Ss}}$ have not been investigated.

Clearly $\pi\chi A \leq \pi A$ for any infinite BA A. The difference between $\pi\chi$ and π can be large, for example in a finite-cofinite algebra: as in the proof of Theorem 9.1, $\pi\chi A = \omega$ for a finite-cofinite algebra A. $\pi\chi A > \mathrm{d}A$ for some free algebras A; a free algebra also shows that $\pi\chi A$ can be greater than Length$A$. $\pi\chi A >$ Ind$A$ for A the interval algebra on an uncountable cardinal κ, and Depth$A > \pi\chi A$ for A the interval algebra on $\omega^* \cdot (\kappa + 1)$; both of these results are clear on the basis of the description of $\pi\chi$ for interval algebras given at the end of this section.

There are two interesting positive results concerning the relationship of $\pi\chi$ with our earlier cardinal functions. The first of these is true for arbitrary non-discrete regular Hausdorff spaces, with no complications in the proof from the BA case:

Theorem 9.6. $\mathrm{d}X \leq \pi\chi X^{\mathrm{c}X}$ *for any non-discrete regular Hausdorff space* X.

Proof. By non-discreteness, $\pi\chi X \geq \omega$; this is easy to check, following the lines of the argument at the beginning of this section. For each $x \in X$ let $\mathcal{O}_x$ be a family of non-empty open subsets of X such that $|\mathcal{O}_x| \leq \pi\chi X$ and for every neighborhood U of x there is a $V \in \mathcal{O}$ such that $V \subseteq U$. Now we define subsets $Y_\alpha \subseteq X$ and collections $\mathcal{P}_\alpha$ of open sets for $\alpha < (\mathrm{c}X)^+$ by induction so that the following conditions hold:

(1) $|Y_\alpha| \leq (\pi\chi X)^{\mathrm{c}X}$;

(2) $|\mathcal{P}_\alpha| \leq (\pi\chi X)^{\mathrm{c}X}$.

Fix $x_0 \in X$. Set $Y_0 = \{x_0\}$ and $\mathcal{P}_0 = \mathcal{O}_{x_0}$. Suppose that Y_β and $\mathcal{P}_\beta$ have been defined for all $\beta < \alpha$. If α is a limit ordinal, set $Y_\alpha = \bigcup_{\beta<\alpha} Y_\beta$ and $\mathcal{P}_\alpha = \bigcup_{\beta<\alpha} \mathcal{P}_\beta$. Now suppose that α is a successor ordinal $\beta + 1$. Set

$$Q_\alpha = \{\mathcal{R} : \mathcal{R} \subseteq \mathcal{P}_\beta, |\mathcal{R}| \leq \mathrm{c}X, \overline{\bigcup \mathcal{R}} \neq X\}$$

Clearly $|Q_\alpha| \leq \pi\chi X^{\mathrm{c}X}$. For every $\mathcal{R} \in Q_\alpha$ choose $\phi_\mathcal{R} \in X \backslash \overline{\bigcup \mathcal{R}}$ and put

$$Y_\alpha = Y_\beta \cup \{\phi_\mathcal{R} : \mathcal{R} \in Q_\alpha\},$$

$$\mathcal{P}_\alpha = \bigcup_{x \in Y_\alpha} \mathcal{O}_x.$$

This finishes the definition. Now we claim

(3) $L \stackrel{\text{def}}{=} \bigcup_{\alpha<(\mathrm{c}X)^+} Y_\alpha$ is dense in X.

Since $|L| \leq (\pi\chi X)^{\mathrm{c}X}$, (3) finishes the proof. To prove (3), suppose that it is not true. Then by regularity, there is an open U such that $\overline{L} \subseteq U \subseteq \overline{U} \neq X$.

Set $\mathcal{P}^* = \bigcup_{x\in L}\mathcal{O}_x$, and $\mathcal{T} = \{V \in \mathcal{P}^* : V \subseteq U\}$. Let $\mathcal{R}$ be a maximal disjoint subset of $\mathcal{T}$. Then $L \subseteq \overline{\bigcup\mathcal{R}}$; for, if $x \in L\backslash\overline{\bigcup\mathcal{R}}$, then $x \in U\backslash\overline{\bigcup\mathcal{R}}$, which is open, so there is a $V \in \mathcal{O}_x$ such that $V \subseteq U\backslash\overline{\bigcup\mathcal{R}}$, and $\mathcal{R} \cup \{V\}$ contradicts the maximality of $\mathcal{R}$. Also, $\overline{\bigcup\mathcal{R}} \subseteq \overline{\bigcup\mathcal{T}} \subseteq \overline{U} \neq X$. Since $\mathcal{R} \subseteq \mathcal{P}_\beta$ for some $\beta < (\mathrm{c}X)^+$, it follows that $\mathcal{R} \in Q_\beta$ for some $\beta < (\mathrm{c}X)^+$, and hence we get $\phi_\mathcal{R} \in X\backslash\overline{\bigcup\mathcal{R}} \subseteq X\backslash L$, contradiction. □

Theorem 9.7. $\mathrm{d}A \cdot \pi\chi A = \pi A$ *for any infinite BA* A.

PROOF. We already know that $\mathrm{d}A \le \pi A$ and $\pi\chi A \le \pi A$. Now let D be a dense subset of Ult$A$ with $|D| = \mathrm{d}A$, and for each $F \in D$ let X_F be a local base for F of size $\le \pi\chi A$. Clearly $\bigcup_{F\in D} X_F$ is dense in A, as desired. □

The functions $\pi\chi_{\mathrm{Sr}}$ and $\pi\chi_{\mathrm{Hr}}$ have not been investigated.

Concerning $\pi\chi$ for special classes of algebras, we first give a description of what happens for interval algebras. Let L be a linearly ordered set with first element 0, and let A be the interval algebra on L. The ultrafilters on A are in one-one correspondence with the final segments of L; corresponding to the ultrafilter F is the segment $\{a \in L : [0, a) \in F\}$. Given a terminal segment T of L, let κ be the type of a shortest cofinal sequence in $L\backslash T$ and λ the type of a shortest coinitial sequence in T. If both κ and λ are infinite, then $\pi\chi F$ is the minimum of κ and λ. If one is infinite and the other is 1, then $\pi\chi F$ is the infinite one. If both are 1, then $\pi\chi F$ is 1. From this description it is easy to construct a linear order L such that if A is the interval algebra on L then $\pi\chi A < \chi A$: for example, let L be $0 + \omega^* \cdot \omega_1 + \omega^*$. The above description implies that $\pi\chi A = \omega$, while if F is the ultrafilter corresponding to the terminal segment ω^*, then $\chi F = \omega_1$. In this example we also have $\pi\chi A < \mathrm{Depth}A$. The description of $\pi\chi$ also shows that $\pi\chi A \le \mathrm{Depth}A$ for an interval algebra A.

If A is complete, then $\mathrm{c}A \le \pi\chi A$: in fact, suppose that $\pi\chi A < \mathrm{c}A$. Let X be disjoint in A with $\sum X = 1$ and $|X| = (\pi\chi A)^+$. Let F be an ultrafilter on A such that $\sum(X\backslash Y) \in F$ for each $Y \subset X$ such that $|Y| < |X|$. Let Y be a π-base for F with $|Y| < |X|$. For each $y \in Y$ choose $x_y \in X$ such that $y \cdot x_y \neq 0$. Then $\{x_y : y \in Y\}$ is a π-base for $F \cap \langle X\rangle^{\mathrm{cm}}$ (the complete subalgebra of A generated by X). But $-\sum_{y\in Y} x_y \in F \cap \langle X\rangle^{\mathrm{cm}}$, contradiction.

We do not know whether $\pi A = \pi\chi A$ for A complete.

Problem 27. *Is* $\pi A = \pi\chi A$ *for* A *complete?*

In section 4 we gave an example of a complete algebra A with the property that $\mathrm{d}A < \pi A$; hence by Theorem 9.7 we have $\mathrm{d}A < \pi\chi A$ also.

10. TIGHTNESS

Again we note first of all that if F is a non-principal ultrafilter in a BA A, then $\mathrm{t}F \geq \omega$. To see this, note that for each $x \in F$ there is a $y \notin F$ such that $0 < y < x$; hence there is an ultrafilter G_x such that $x \in G_x$ but $G_x \neq F$. Let $Y = \{G_x : x \in F\}$. Thus $F \subseteq \bigcup Y$. Suppose that Z is a finite subset of Y such that $F \subseteq \bigcup Z$. But it is a very elementary exercise to show that no ultrafilter is included in a finite union of other, different, ultrafilters. So, $\mathrm{t}F \geq \omega$, and hence $\mathrm{t}A \geq \omega$ for every infinite BA A.

From at the definition of tightness it is clear that $\mathrm{t}(A \times B) = \max\{\mathrm{t}A, \mathrm{t}B\}$. Furthermore, $\mathrm{t}(\prod^{\mathrm{w}}_{i\in I} A_i) = \sup_{i\in I}\mathrm{t}A_i$ for any system $\langle A_i : i \in I\rangle$ of non-trivial BA's with I infinite. By the topological description of weak products, to prove this it suffices to show that $\mathrm{t}F = \omega$ for the "new" ultrafilter $F \stackrel{\text{def}}{=} \{x \in \prod^{\mathrm{w}}_{i\in I} A_i :$ there is a finite subset F of I such that $x_i = 1$ for all $i \in I\backslash F\}$. To see this, let J be any infinite subset of I, and for each finite subset X of J let x_X be the function which is 1 on $I\backslash X$ and 0 on X itself. Suppose that $F \subseteq \bigcup Y$, where $Y \subseteq \mathrm{Ult}A$. For each finite subset X of J let G_X be a member of Y such that $x_X \in G_X$, and let $Z = \{G_X : X$ is a finite subset of $J\}$. Clearly $F \subseteq \bigcup Z$, as desired.

Note that this argument again shows that tightness is attained in $\prod^{\mathrm{w}}_{i\in I} A_i$ iff there is an $i \in I$ such that $\mathrm{t}\prod^{\mathrm{w}}_{i\in I} A_i = \mathrm{t}A_i$ and tightness is attained in A_i (for infinite A_i's). From this, the attainment property of tightness follows: for each limit cardinal $\kappa > \omega$ there is a BA A with tightness κ not attained: take the weak product of $\langle A_\lambda : \omega < \lambda < \kappa, \lambda$ a cardinal$\rangle$, where A_λ is the free BA of size λ.

It is appropriate at this point to recall the equivalent to tightness involving free sequences; see Theorem 2.10. There is an algebraic equivalent of free sequences. Let A be a BA. A *free sequence* in A is a sequence $\langle x_\xi : \xi < \alpha\rangle$ of elements of A such that if $\xi < \alpha$ and F and G are finite subsets of ξ and $\alpha\backslash\xi$ respectively, then $\prod_{\eta\in F} x_\eta \cdot \prod_{\eta\in G} -x_\eta \neq 0$. Then A has a free sequence of length α iff $\mathrm{Ult}A$ has a free sequence (in the topological sense, defined in section 2) of length α. In fact, first suppose that $\langle x_\xi : \xi < \alpha\rangle$ is a free sequence in A. For each $\xi < \alpha$ let F_ξ be an ultrafilter containing $\{x_\eta : \eta \leq \xi\} \cup \{-x_\eta : \xi < \eta < \alpha\}$. This is possible by the definition above. It is easy to check that $\langle F_\xi : \xi < \alpha\rangle$ is a free sequence in $\mathrm{Ult}A$. Conversely, let $\langle F_\xi : \xi < \alpha\rangle$ be a free sequence in $\mathrm{Ult}A$. Then by the definition of free sequences in spaces, for each $\xi < \alpha$ there is a $x_\xi \in A$ such that $\{F_\eta : \eta < \xi\} \subseteq \mathcal{S}(-x_\xi)$ and $\{F_\eta : \xi \leq \eta\} \subseteq \mathcal{S}x_\xi$. Then $\langle x_\xi : \xi < \alpha\rangle$ is a free sequence in A. This equivalence shows, in particular, that $\mathrm{Ind}A \leq \mathrm{t}A$. Note that tightness in these two free sequence senses have the same attainment properties: one is attained iff the other is.

Concerning attainment in the free sequence sense, we first show

Theorem 10.1. *If κ is an infinite cardinal with $\mathrm{cf}\kappa > \omega$ and $\langle A_i : i \in I\rangle$*

is a system of BA's none of which has a free sequence of type κ, then also $\prod^{\mathrm{w}}_{i\in I} A_i$ does not have a free sequence of type κ.

PROOF. Suppose that $\langle F_\alpha : \alpha < \kappa\rangle$ is a free sequence in $\mathrm{Ult}\prod^{\mathrm{w}}_{i\in I} A_i$. We think of $\mathrm{Ult}\prod^{\mathrm{w}}_{i\in I} A_i$ as the one-point compactification of the disjoint union of all of the spaces $\mathrm{Ult}A_i$. We may assume that the "new" ultrafilter G is not among the F_α's. For each $\alpha < \kappa$ let iF_α be the unique $i \in I$ such that $F_\alpha \in \mathrm{Ult}A_i$. Set $J = \{iF_\alpha : \alpha < \kappa\}$. Then $|J| \geq \mathrm{cf}\kappa$, since $\kappa = \bigcup_{j\in J}\{\eta < \kappa : iF_\eta = j\}$. Now $J = \bigcup_{\xi<\kappa}\{iF_\eta : \eta < \xi\}$, so it follows from $\mathrm{cf}\kappa > \omega$ that there is a $\xi < \kappa$ such that $\{iF_\eta : \eta < \xi\}$ is infinite. Clearly $|\{iF_\eta : \xi \leq \eta\}| \geq \mathrm{cf}\kappa$ by the above argument, so it follows that

$$G \in \overline{\{F_\eta : \eta < \xi\}} \cap \overline{\{F_\eta : \xi \leq \eta < \kappa\}},$$

which contradicts the free sequence property. □

It follows from Theorem 10.1 that for every κ with $\mathrm{cf}\kappa > \omega$ there is a BA with tightness κ not attained in the free sequence sense.

Now we turn to the case of cofinality ω:

Theorem 10.2. *Let $\mathrm{t}A = \kappa$, where κ is a singular cardinal of cofinality ω. Then A has a free sequence of length κ.*

PROOF. This will be a modification of the proof of 2.2; see also Theorem 2.11. An element $a \in A$ is called a *μ-element* if for some ideal I of $A \restriction a$, the algebra $(A \restriction a)/I$ has a strictly increasing sequence of type μ. Let $\langle \lambda_i : i < \omega\rangle$ be a strictly increasing sequence of infinite regular cardinals with supremum κ. We call an element $a \in A$ an *∞-element* if it is a λ_i-element for all $i < \omega$.

(1) If a is an ∞-element and $a = b + c$ with $b \cdot c = 0$, then b is an ∞-element or c is an ∞-element.

For, it is enough to show that for every $i < \omega$, either b is a λ_i-element or c is a λ_i-element. Suppose that for some $i < \omega$, neither b nor c is a λ_i-element. Let I be an ideal in $A \restriction a$ and $\langle [x_\alpha] : \alpha < \lambda_i\rangle$ a strictly increasing sequence of elements in $(A \restriction a)/I$. Now if $\alpha < \beta < \lambda_i$, then

$$x_\alpha \cdot b \cdot -(x_\beta \cdot b) = x_\alpha \cdot -x_\beta \cdot b \in I \cap (A \restriction b),$$

and hence in $A \restriction b$ we have $[x_\alpha \cdot b] \leq [x_\beta \cdot b]$. Hence there is an $\alpha < \lambda_i$ such that if $\alpha < \beta < \gamma < \lambda_i$ then $x_\gamma \cdot -x_\beta \cdot b \in I$. Similarly for c: there is an $\alpha' < \lambda_i$ such that if $\alpha' < \beta < \gamma < \lambda_i$, then $x_\gamma \cdot -x_\beta \cdot c \in I$. But then if $\max(\alpha, \alpha') < \beta < \gamma < \lambda_i$ we get $x_\gamma \cdot -x_\beta \in I$, contradiction. This proves (1).

Now we construct disjoint elements $a_0, a_1, \ldots$ such that a_i is a λ_i-element for all $i < \omega$. Suppose that a_i has been constructed for all $i < n$ so that

$\prod_{i<n} -a_i$ is an ∞-element. Now there exists an ideal I in $A \upharpoonright \prod_{i<n} -a_i$ with a sequence $\langle [x_\alpha] : \alpha < \lambda_{n+1} \rangle$ strictly increasing in $(A \upharpoonright \prod_{i<n} -a_i)/I$. Then clearly

(2) x_{λ_n} is a λ_n-element.

Now by (1) and (2) there is a λ_n-element a_n such that $\prod_{i\leq n} -a_i$ is an ∞-element.

Now for each $i < \omega$ choose an ideal I_i in $A \upharpoonright a_i$ such that $(A \upharpoonright a_i)/I_i$ has a chain of type λ_i. Let $J = \langle \bigcup_{i<\omega} I_i \rangle^{Id}$. Then $J \cap (A \upharpoonright a_i) = I_i$ for each $i < \omega$, and hence A/J has a chain of type λ_i for all $i < \omega$. Hence as in the proof of 2.2, A/J has a chain of type κ, as desired (see the proof of 2.11). □

We also recall from Theorem 9.5 two more equivalents of tightness: $\mathrm{t}A = \pi\chi_{\mathrm{H}+}A = \pi\chi_{\mathrm{h}+}A$. And, as mentioned after the proof of Theorem 9.5, $\pi\chi_{\mathrm{H}+}$ and $\pi\chi_{\mathrm{h}+}$ have the same attainment properties, while $\pi\chi_{\mathrm{H}+}$ attained implies that t is attained in the free sequence sense. We have now mentioned all that we know about the relationship between attainment for the various equivalents of tightness. The following questions remain open.

Problem 28. *Does attainment of tightness imply attainment in the free sequence sense?*

Problem 29. *For* $\mathrm{cf}\kappa > \omega$, *does attainment of tightness in the free sequence sense imply attainment of tightness in the defined sense?*

Problem 30 *Does attainment of tightness in the free sequence sense imply attainment of* $\pi\chi_{\mathrm{H}+}$*?*

We return to the discussion of products.

Theorem 10.3. *If* $\langle A_i : i \in I \rangle$ *is a system of non-trivial BA's, with* I *infinite, then* $\mathrm{t}(\prod_{i\in I} A_i) \geq \max(2^{|I|}, \sup_{i\in I}\mathrm{t}A_i)$.

PROOF. Let $i \in I$; we show that $\mathrm{t}A_i \leq \mathrm{t}(\prod_{i\in I} A_i)$. For any ultrafilter F on A_i, let $F^+ = \{x \in \prod_{i\in I} A_i : x_i \in F\}$; clearly F^+ is an ultrafilter on $\prod_{i\in I} A_i$. It suffices to show that $\mathrm{t}F \leq \mathrm{t}F^+$. Suppose that $F \subseteq \bigcup Y$, where $Y \subseteq \mathrm{Ult}A_i$. Then clearly $F^+ \subseteq \bigcup\{H^+ : H \in Y\}$. So, let Z be a subset of Y of power $\leq \mathrm{t}F^+$ such that $F^+ \subseteq \bigcup\{H^+ : H \in Z\}$. Then $F \subseteq \bigcup Z$, as is easily verified. So $\mathrm{t}F \leq \mathrm{t}F^+$, as desired.

Since independence is less than or equal to tightness, it also follows that $2^{|I|} \leq \mathrm{t}(\prod_{i\in I} A_i)$. □

Again, Example 8.5 can be used to show that tightness can jump in a product. To prove this, it suffices to show that $\mathrm{t}B_n = \omega$ for each positive integer n, where B_n is defined in 8.5. So, let F be any ultrafilter on B_n, and suppose $F \subseteq \bigcup Y$, where $Y \subseteq \text{Ult}B_n$. From the description of elements given in 8.5, we know that $\{\alpha < \kappa : a^n_\alpha \in H\}$ is finite, for any ultrafilter H on B_n. Again, let $M = \{\alpha < \kappa : a^n_\alpha \in F\}$. Now we define a sequence $\langle G_n : n < \omega\rangle$ of members of Y by induction. Let G_0 be any member of Y. Suppose that G_m has been defined. Since $\prod_{\alpha\in M} a^n_\alpha \cdot \prod\{-a^n_\alpha : \alpha \notin M \text{ and } a^n_\alpha \in G_i \text{ for some } i \leq m\} \in F$, we can choose $G_{m+1} \in Y$ containing this element. Now we claim that $F \subseteq \bigcup_{m\in\omega} G_m$. It suffices to show that every element of F of the form

$$\prod_{\alpha\in M} a^n_\alpha \cdot \prod_{\alpha\in N} -a^n_\alpha$$

is in $\bigcup_{m\in\omega} G_m$, where N is any finite subset of $\kappa\backslash M$. Now we define $P = \{\alpha \in N : a^n_\alpha \in G_m \text{ for some } m \in \omega\}$. For each $\alpha \in P$ choose $m(\alpha) \in \omega$ such that $a^n_\alpha \in G_{m(\alpha)}$. Let m be greater than all of the elements $m(\alpha)$, $\alpha \in P$. Clearly the above element is in G_m, by construction.

Note that there can be superatomic interval algebras with high tightness; this is clear from Theorem 2.11.

The behaviour of tightness with respect to ultraproducts has not been investigated.

It is easy to verify that if A is a homomorphic image of B, then $\text{t}A \leq \text{t}B$. This is also true if A is a subalgebra of B: suppose that $F \in \text{Ult}A$, $Y \subseteq \text{Ult}A$, and $F \subseteq \bigcup Y$. Let $D = \{G \in \text{Ult}B : G \cap A \in Y\}$. Choose G maximal in $\{H : H \text{ is a filter on } B, \text{ and } F \subseteq H \subseteq \bigcup D\}$. We claim that G is an ultrafilter. Suppose $b \in B\backslash G$. Then $\langle G \cup \{b\}\rangle^{Fi} \not\subseteq \bigcup D$; say $c \in G$ and $c \cdot b \notin \bigcup D$. If also $-b \notin G$, we get $d \in G$ with $d \cdot -b \notin \bigcup D$. Say $c \cdot d \in H \in D$. Then $c \cdot d \cdot b \notin H$ and $c \cdot d \cdot -b \notin H$, contradiction. Thus, as claimed, G is an ultrafilter.

Choose $E \subseteq D$ such that $|E| \leq \text{t}B$ and $G \subseteq \bigcup E$. Then we have $F \subseteq \bigcup\{H \cap B : H \in E\}$, as desired.

The tightness of free products is described by a theorem of Malyhin [72]; we give the result here. It depends on the following lemma.

Lemma 10.4. *Let f be a closed continuous mapping from X onto Y. Suppose $A \subseteq X$, $y \in Y$, and $f^{-1}[\{y\}] \cap \overline{A} \neq 0$. Then there is a $B \in [A]^{\leq \text{t}Y}$ such that $f^{-1}[\{y\}] \cap \overline{B} \neq 0$.*

PROOF. Choose $x \in f^{-1}[\{y\}] \cap \overline{A}$. Now

(1) $y \in \overline{f[A]}$.

To prove (1), let $y \in U$, where the set U is open. Thus $f^{-1}[\{y\}] \subseteq f^{-1}[U]$, so $x \in f^{-1}[U] \cap \overline{A}$, and it follows that $f^{-1}[U] \cap A \neq 0$; say $a \in f^{-1}[U] \cap A$. Then $fa \in U \cap f[A]$, as desired in (1).

Choose $B \in [A]^{\leq \text{t}Y}$ such that $y \in \overline{f[B]}$. We claim that $y \in f[\overline{B}]$ (as desired). For, $f[\overline{B}]$ is closed; if $y \notin f[\overline{B}]$, then $y \in X\backslash f[\overline{B}]$, which is open, so $(X\backslash f[\overline{B}]) \cap f[B] \neq 0$, contradiction. □

Theorem 10.5. *If X and Y are Hausdorff spaces and Y is compact, then* $\mathrm{t}(X \times Y) = \max(\mathrm{t}X, \mathrm{t}Y)$

PROOF. We have $\geq$ by an easy argument. Now let $\kappa = \max(\mathrm{t}X, \mathrm{t}Y)$. Suppose that $(x, y) \in \overline{A}$; we want to find $B \in [A]^{\leq\kappa}$ such that $(x, y) \in \overline{B}$. Let π_1 be the projection into the first coordinate X; we know that π_1 is closed and continuous. For every open U with $y \in U$ we have

$$(x, y) \in \overline{A \cap (X \times U)},$$

i.e.,

$$\pi_1^{-1}[\{x\}] \cap \overline{A \cap (X \times U)} \neq 0.$$

By the lemma, choose $B_U \in [A \cap (X \times U)]^{\leq \mathrm{t}X}$ such that $\pi_1^{-1}\{x\} \cap \overline{B}_U \neq 0$, and choose $z_U \in Y$ with $(x, z_U) \in \overline{B}_U$. Let

$$C = \{z_U : U \text{ is open and } y \in U\}.$$

We claim

(1) $y \in \overline{C}$.

For, let $y \in W$, where W is open. Choose U open with $y \in U \subseteq \overline{U} \subseteq W$. Note that $(x, z_U) \in \overline{B}_U \subseteq \overline{X \times U}$, so $z_U \in \overline{U} \subseteq W$, as desired.

Hence we can choose $\mathcal{B} \in [\{U : U \text{ is open and } y \in U\}]^{\leq \mathrm{t}Y}$ such that $y \in \overline{\{z_U : U \in \mathcal{B}\}}$. We claim

(2) $(x, y) \in \overline{\bigcup_{U \in \mathcal{B}} B_U}$ (as desired).

For, let $(x, y) \in V \times W$, where V and W are open. Choose $U \in \mathcal{B}$ with $z_U \in W$. Thus $(x, z_U) \in V \times W$, and $(x, z_U) \in \overline{B}_U$, so $(V \times W) \cap B_U \neq 0$, as desired. □

Theorem 10.6. *If each space X_i is compact Hausdorff with at least 2 elements, then* $\mathrm{t}(\prod_{i \in I} X_i) = \max(|I|, \sup_{i \in I} \mathrm{t}X_i)$.

PROOF. For $\geq$, by an easy argument $\mathrm{t}X_i \leq \mathrm{t}(\prod_{i \in I} X_i)$ for every $i \in I$. To see that also $|I| \leq \mathrm{t}(\prod_{i \in I} X_i)$, we proceed as follows. Let x and z be members of $\prod_{i \in I} X_i$ such that $x_i \neq z_i$ for each $i \in I$. For each $F \in [I]^{<\omega}$ let y^F be the member of $\prod_{i \in I} X_i$ such that $y_i^F = x_i$ for all $i \in F$, while $y_i^F = z_i$ for all $i \in I \backslash F$. Clearly $x \in \overline{\{y^F : F \in [I]^{<\omega}\}}$. Suppose that $J \in [[I]^{<\omega}]^{<|I|}$ and $x \in \overline{\{y^F : F \in J\}}$. Then there is an $i \in I \backslash \bigcup J$. Let U be an open neighborhood of x_i such that $z_i \notin U$, and let $V = \{w \in \prod_{i \in I} X_i : w_i \in U\}$. Thus $x \in V$, V is open in $\prod_{i \in I} X_i$, but $y^F \notin V$ for all $F \in J$, contradiction.

For $\mathrm{t}(\prod_{i \in I} X_i) \leq \max(|I|, \sup_{i \in I} \mathrm{t}X_i)$, let $\kappa = \sup_{i \in I} \mathrm{t}X_i$. Suppose also that $A \subseteq \prod_{i \in I} X_i$, $x \in \prod_{i \in I} X_i$, and $x \in \overline{A}$. For every $F \in [I]^{<\omega}$ we then have that $x \restriction F \in \overline{\{y \restriction F : y \in A\}}$, so by Theorem 10.4 it is possible to choose

$B_F \in [A]^{\leq\kappa}$ such that $x \restriction F \in \overline{\{y \restriction F : y \in B_F\}}$. Then it is easily checked that $x \in \overline{\bigcup\{B_F : F \in [I]^{<\omega}\}}$, as desired. □

The behaviour of tightness in the free sequence sense under unions of chains of BA's is similar to the case of cellularity (Theorem 1.1). The definition of ordinary sup-function does not quite fit, but essentially the same proof can be used:

Theorem 10.7. *Let κ and λ be infinite cardinals, with λ regular. Then the following conditions are equivalent:*

(i) $\mathrm{cf}\kappa = \lambda$.

(ii) There is a stictly increasing sequence $\langle A_\alpha : \alpha < \kappa\rangle$ of Boolean algebras each with no free sequence of type κ such that $\bigcup_{\alpha<\lambda} A_\alpha$ has a free sequence of type κ. □

In view of the equivalence of tightness with its free sequence variant, 10.7 also applies to tightness when κ is a successor cardinal. If we try to go through the proof of Theorem 1.1 for tightness, we find that *(i)* $\Rightarrow$ *(ii)* works, by a suitable example, while *(ii)* at least implies that $\mathrm{cf}\kappa \leq \lambda$, by an easy argument. The remaining possiblity is open:

Problem 31. *Do there exist cardinals κ and λ such that λ is regular, κ is singular, $\mathrm{cf}\kappa < \lambda$, and there is a sequence $\langle A_\alpha : \alpha < \kappa\rangle$ of BA's each with tightness less than κ such that $\bigcup_{\alpha<\lambda} A_\alpha$ has tightness κ?*

We turn to derived functions for tightness. By Theorem 9.5 we have that $\mathrm{t}_{\mathrm{H}+} = \mathrm{t}_{\mathrm{h}+} = \mathrm{t}A$. Clearly $\mathrm{t}_{\mathrm{S}+} = \mathrm{t}A$, $\mathrm{t}_{\mathrm{S}-}A = \omega$, and ${}_{\mathrm{d}}\mathrm{t}_{\mathrm{S}+}A = \mathrm{t}A$. We do not know the status of $\mathrm{t}_{\mathrm{H}-}$; a particular problem is

Problem 32. *Is* $\mathrm{t}_{\mathrm{H}-}A = \mathrm{Card}_{\mathrm{H}-}A$?

Note that ${}_{\mathrm{d}}\mathrm{t}_{\mathrm{S}-}A \neq \mathrm{t}A$ in general; this can be seen by considering $\mathcal{P}\omega$ and its dense subalgebra consisting of the finite and cofinite subsets of ω.

Recall also our earlier results that $\mathrm{Depth}_{\mathrm{H}+}A = \mathrm{t}A = \pi\chi_{\mathrm{H}+}A$; see Theorems 2.11, 9.5.

One can define another cardinal function related to tightness as follows: $\mathrm{t}_{\mathrm{inf}}A = \inf\{\mathrm{t}F : F \in \mathrm{Ult}A\}$; this function has not been investigated.

Next we mention more about the relationships between tightness and our previously introduced functions. By Theorem 2.11 we have $\mathrm{Depth}A \leq \mathrm{t}A$ for any BA A; the difference can be big, for example in a free algebra. $\pi\chi A \leq \mathrm{t}A$ by Theorem 9.5; we do not know whether the difference can be large. An example where they differ is $\mathcal{P}\kappa$: $\pi\chi\mathcal{P}\kappa \geq \kappa$ by Theorem 9.2, while $\pi\chi\mathcal{P}\kappa \leq \pi\mathcal{P}\kappa = \kappa$; so $\pi\chi\mathcal{P}\kappa = \kappa$, and clearly $\mathrm{t}\mathcal{P}\kappa = 2^\kappa$.

Problem 33. *Can the difference between $\pi\chi$ and* t *be large?*

We observed at the beginning of this section that $\mathrm{Ind}A \leq \mathrm{t}A$; the difference is large in some interval algebras. Obviously $\mathrm{t}A \leq |A|$. We do not know whether it is possible to have $\mathrm{t}A > \mathrm{Irr}A$.

Problem 34. *Is it possible to have* $\mathrm{t}A > \mathrm{Irr}A$?

$\mathrm{t}A > \pi A$ for $A = \mathcal{P}\omega$. $\mathrm{t}A > \mathrm{Length}A$ for A an uncountable free BA. $\mathrm{Length}A > \mathrm{t}A$ for A the interval algebra on the reals. $\mathrm{c}A > \mathrm{t}A$ for A an uncountable finite-cofinite algebra.

Concerning tightness for special classes of algebras, we mention only that $\mathrm{t}A = |A|$ whenever A is complete.

11. SPREAD

The following theorem gives some equivalent definitions of spread.

Theorem 11.1. *For any infinite BA A, $\mathrm{s}A$ is equal to each of the following cardinals:*

$\sup\{|X| : X$ *is a minimal set of generators of* $\langle X\rangle^{\mathrm{Id}}\}$;
$\sup\{|X| : X$ *is ideal-independent*$\}$;
$\sup\{|X| : X$ *is the set of all atoms in some homomorphic image of* $A\}$;
$\sup\{|\mathrm{At}B| : B$ *is an atomic homomorphic image of* $A\}$;
$\sup\{\mathrm{c}B : B$ *is a homomorphic image of* $A\}$.

PROOF. Six cardinals are mentioned in this theorem; let them be denoted by $\kappa_0, \ldots \kappa_5$ in the order that they are mentioned. In Theorem 1.2 we proved that $\kappa_0 = \kappa_2$, and in Theorem 1.3 that $\kappa_2 = \kappa_5$. It is obvious that $\kappa_1 = \kappa_2$. To show that $\kappa_3 \leq \kappa_4$, suppose that B is a homomorphic image of A with an infinite number of atoms. Let I be the ideal $\langle\{x : x \cdot a = 0$ for every atom a of $B\}\rangle^{\mathrm{Id}}$ of B. Clearly B/I is atomic with the same number of atoms as B. This shows that $\kappa_3 \leq \kappa_4$. Obviously $\kappa_4 \leq \kappa_5$ Finally, for $\kappa_5 \leq \kappa_3$, let B be a homomorphic image of A, and let D be an infinite disjoint subset of B. We show how to find an atomic homomorphic image C of B with exactly $|D|$ atoms. Let M be the subalgebra of B generated by D. Let f be an extension of the identity on D to a homomorphism of B into $\overline{D}$; the image of B under f is as desired. □

As to attainment of spread, first note that all of the equivalents of spread given in Theorem 11.1 have the same attainment properties. We state the facts known about attainment of spread without proof: (1) Spread is always attained for singular strong limit cardinals: see Juhász [80] Theorem 4.2; (2) Spread is always attained for singular cardinals of cofinality ω; see Juhász [80], Theorem 4.3; (3) Assuming V=L, if κ is inaccessible but not weakly compact, then there is a BA A with spread κ not attained: see Juhász [71], example 6.6; (4) If $\mathrm{s}A$ is weakly compact, then $\mathrm{s}A$ is attained: see Juhász [71], remark following 3.2; (5) If 2^ω is a limit cardinal, then there is a BA A with spread 2^ω not attained; see Corollary 1.9.

An infinite BA A has an infinite disjoint subset D, which gives rise to an infinite discrete subspace of Ult$A$. So $\mathrm{s}A$ is always infinite.

The following theorem is obvious upon looking at its topological dual:

Theorem 11.2. *Suppose that $\langle A_i : i \in I\rangle$ is a system of BA's each with at least two elements. Then* $\mathrm{s}(\prod^{\mathrm{w}}_{i\in I} A_i) = \max(|I|, \sup_{i\in I}\mathrm{s}A_i)$. □

Clearly $\mathrm{s}(\prod_{i\in I} A_i) \geq \max(2^{|I|}, \sup_{i\in I}\mathrm{s}A_i)$; we do not know whether equality holds:

Problem 35. $\mathrm{s}(\prod_{i\in I} A_i) = \max(2^{|I|}, \sup_{i\in I}\mathrm{s}A_i)$?

The spread of ultraproducts has not been investigated.

If A is a subalgebra or a homomorphic image of B, then $\mathrm{s}A \leq \mathrm{s}B$, and the difference can be large.

We turn to the discussion of the spread of free products. The simplest fact is as follows, and the proof is obvious.

Theorem 11.3. *If $\langle A_i : i \in I\rangle$ is a system of BA's each with at least 4 elements, then* $\mathrm{s}(\oplus_{i\in I}A_i) \geq \max(|I|, \sup_{i\in I}\mathrm{s}A_i)$. □

Equality does not hold in Theorem 11.3, in general. For example, let A be the interval algebra on the reals. We observed in Corollary 1.7 that $\mathrm{s}A = \omega$. Here is a system of 2^ω ideal independent elements in $A \oplus A$: for each real number r, let $a_r = [r,\infty) \times (-\infty, r)$ (considered as an element of $A \oplus A$). Suppose that F is a finite subset of $\mathbf{R}$, $r \in \mathbf{R}\backslash F$, and $a_r \in \langle a_s : s \in F\rangle^{\mathrm{Id}}$. Thus

$$[r,\infty) \times (-\infty, r) \cdot \prod_{s\in F}((-\infty, s) + [s,\infty)) = 0.$$

But if $T \overset{\mathrm{def}}{=} \{s \in F : r < s\}$ and $U \overset{\mathrm{def}}{=} F\backslash T$, then

$$[r,\infty) \times (-\infty, r) \cdot \prod_{s\in F}((-\infty, s) + [s,\infty)) \geq$$

$$[r,\infty) \times (-\infty, r) \cdot \prod_{s\in T}(-\infty, s) \cdot \prod_{s\in U}[s,\infty) \neq 0,$$

contradiction.

We can, however, give an upper bound for the spread of a free product, namely $\max(|I|, 2^{\sup_{i\in I}\mathrm{s}A_i})$. This is true because $|B| \leq 2^{\mathrm{s}B}$ for any BA B (see Theorem 11.7 below); so

$$\max(|I|, \sup_{i\in I}|A_i|) \leq \mathrm{s}(\oplus_{i\in I}A_i) \leq \max(|I|, 2^{\sup_{i\in I}\mathrm{s}A_i}).$$

Both equalities here can be attained.

We give now the proof that $|B| \leq 2^{\mathrm{s}B}$ for any BA B. It depends on several other results which are of interest. A *network* for a space X is a collection $\mathcal{N}$ of subsets of X such that every open set in X is a union of members of $\mathcal{N}$ (the members of $\mathcal{N}$ are not assumed to be open).

Theorem 11.4. *For any infinite BA A, $|A| = \min\{|\mathcal{N}| : \mathcal{N}$ is a network for* $\mathrm{Ult}A\}$.

PROOF. Clearly $\geq$ holds. Now suppose that $\mathcal{N}$ is a network for Ult$A$. Let $\mathcal{P}$ be the set of all pairs (C, D) such that $C, D \in \mathcal{N}$ and for some disjoint open sets U and V, $C \subseteq U$ and $D \subseteq V$; and for each $(C, D) \in \mathcal{P}$, choose open sets of this sort — call them U_{CD} and V_{CD}. Then let $\mathcal{W}$ be the closure of the set $\{U_{CD}, V_{CD} : (C, D) \in \mathcal{P}\}$ under $\cap$ and $\cup$. We shall now show that $\{\mathcal{S}a : a \in A\} \subseteq \mathcal{W}$, which will prove $\leq$. So, let $a \in A$. For each $F \in \mathcal{S}a$ and $G \notin \mathcal{S}a$ choose disjoint open sets X, Y such that $F \in X$ and $G \in Y$; then choose $C(F, G), D(F, G) \in \mathcal{N}$ such that $F \in C(F, G) \subseteq X$ and $G \in D(F, G) \subseteq Y$. Thus $(C(F, G), D(F, G)) \in \mathcal{P}$; so in particular $F \in U_{C(F,G)D(F,G)}$ and $G \in V_{C(F,G)D(F,G)}$. Now fix $G \notin \mathcal{S}a$. Thus by compactness of $\mathcal{S}a$ we get a finite subset $\mathcal{F}$ of $\mathcal{S}a$ such that $\mathcal{S}a \subseteq \bigcup_{F \in \mathcal{F}} C(F, G)$. Let $U(G) = \bigcup_{F \in \mathcal{F}} C(F, G)$ and $V(G) = \bigcap_{F \in \mathcal{F}} D(F, G)$. Thus $\mathcal{S}a \subseteq U(G)$ and $G \in V(G)$, and $U(G)$ and $V(G)$ are disjoint. By compactness of Ult$A \backslash \mathcal{S}a$ there is a finite subset $\mathcal{G}$ of Ult$A \backslash \mathcal{S}a$ such that Ult$A \backslash \mathcal{S}a \subseteq \bigcup_{G \in \mathcal{G}} V(G)$. Since also $\mathcal{S}a \subseteq \bigcap_{G \in \mathcal{G}} U(G)$, and $\bigcup_{G \in \mathcal{G}} V(G)$ and $\bigcap_{G \in \mathcal{G}} U(G)$ are disjoint, we have $\mathcal{S}a = \bigcap_{G \in \mathcal{G}} U(G) \in \mathcal{W}$, as desired. □

Theorem 11.5. *For any infinite BA A, $\mathrm{t}A \leq \mathrm{s}A$.*

PROOF. By Theorem 2.10 it suffices to note that if $\langle F_\xi : \xi < \alpha \rangle$ is a free sequence, then $\langle F_\xi : \xi < \alpha \rangle$ is one-one and $\{F_\xi : \xi < \alpha\}$ is discrete. Let $\xi < \alpha$. There exist clopen sets $\mathcal{S}a, \mathcal{S}b$ such that

$$\begin{aligned} \{F_\eta : \eta < \xi\} \cap \mathcal{S}a &= 0, \\ \{F_\eta : \xi \leq \eta < \alpha\} &\subseteq \mathcal{S}a, \\ \{F_\eta : \eta < \xi + 1 < \alpha\} &\subseteq \mathcal{S}b, \\ \{F_\eta : \xi + 1 \leq \eta < \alpha\} \cap \mathcal{S}b &= 0. \end{aligned}$$

Clearly then $\mathcal{S}(a \cdot b) \cap \{F_\eta : \eta < \alpha\} = \{F_\xi\}$, as desired. □

Lemma 11.6. *If X is a Hausdorff space and $2^\kappa < |X|$, then there is a sequence $\langle F_\alpha : \alpha < \kappa^+ \rangle$ of closed subsets of X such that $\alpha < \beta$ implies $F_\beta \subset F_\alpha$.*

PROOF. For each $f \in \bigcup_{\alpha < \kappa^+} {}^\alpha 2$ we define a closed subset X_f of X. Let $X_0 = X$. For dom$f$ limit, let $X_f = \bigcap_{\alpha < \mathrm{dom} f} X_{f \restriction \alpha}$. Now suppose that X_f has been constructed. If $|X_f| \leq 1$, let $X_{f \frown \langle 0 \rangle} = X_{f \frown \langle 1 \rangle} = X_f$. Otherwise, let $X_{f \frown \langle 0 \rangle}$ and $X_{f \frown \langle 1 \rangle}$ be two proper closed subsets of X_f whose union is X_f. This finishes the construction. Clearly $\bigcup_{\mathrm{dom} f = \alpha} X_f = X$ for all $\alpha < \kappa^+$. Now

(*) there is an $f \in {}^{\kappa^+}2$ such that $|X_{f \restriction \alpha}| \geq 2$ for all $\alpha < \kappa^+$.

For, otherwise, for all $x \in X$ there is an $f \in \bigcup_{\alpha < \kappa^+} {}^\alpha 2$ such that $X_f = \{x\}$, and so

$$|X| \leq \left| \bigcup_{\alpha < \kappa^+} {}^\alpha 2 \right| = 2^\kappa,$$

contradiction. So (*) holds, and it clearly gives the desired result. □

Theorem 11.7. $|B| \leq 2^{\mathrm{s}B}$ *for any BA B.*

PROOF. To start with, we prove:

(1) $\mathrm{d}A \leq 2^{\mathrm{s}A}$.

In fact, suppose that (1) fails. Note that for every $Y \subseteq \mathrm{Ult}A$ of power $< \mathrm{d}A$ we have $\overline{Y} \neq \mathrm{Ult}A$. Hence one can construct two sequences $\langle F_\alpha : \alpha < (2^{\mathrm{s}A})^+\rangle$ and $\langle a_\alpha : \alpha < (2^{\mathrm{s}A})^+\rangle$ such that $a_\alpha \in F_\alpha \in \mathrm{Ult}A$ and $\mathcal{S}a_\alpha \cap \{F_\beta : \beta < \alpha\} = 0$ for all $\alpha < (2^{\mathrm{s}A})^+$. Let $X = \{F_\alpha : \alpha < (2^{\mathrm{s}A})^+\}$. Clearly F is one-one, so $|X| > 2^{\mathrm{s}A}$. By Lemma 11.6, let $\langle K_\alpha : \alpha < (\mathrm{s}A)^+\rangle$ be a system of closed subsets of X such that $\alpha < \beta$ implies that $K_\beta \subset K_\alpha$. Say $F_{\beta_\alpha} \in K_\alpha \backslash K_{\alpha+1}$ for all $\alpha < (\mathrm{s}A)^+$, and choose $b_\alpha \in A$ so that $F_{\beta_\alpha} \in \mathcal{S}b_\alpha \cap X \subseteq X \backslash K_{\alpha+1}$. Then

(2) $\mathcal{S}b_\alpha \cap \{F_{\beta_\gamma} : \gamma > \alpha\} = 0$.

For, suppose $\gamma > \alpha$ and $F_{\beta_\gamma} \in \mathcal{S}b_\alpha$. But $F_{\beta_\gamma} \in K_\gamma \subseteq K_{\alpha+1}$, contradiction.

Define $f : [(\mathrm{s}A)^+]^2 \to 2$ as follows: $f\{\gamma, \delta\} = 0$ iff when $\gamma < \delta$ we have $\beta_\gamma > \beta_\delta$. We now use the partition relation $\mu^+ \to (\omega, \mu^+)$. Since there is no infinite decreasing sequence of ordinals, we get a subset Γ of $(\mathrm{s}A)^+$ of size $(\mathrm{s}A)^+$ such that if $\gamma, \delta \in \Gamma$ and $\gamma < \delta$, then $\beta_\gamma < \beta_\delta$. Hence for any $\alpha \in \Gamma$ we have

$$\mathcal{S}(a_{\beta_\alpha} \cdot b_\alpha) \cap \{F_{\beta_\gamma} : \gamma \in \Gamma\} = \{F_{\beta_\alpha}\},$$

and $\{F_{\beta_\gamma} : \gamma \in \Gamma\}$ is discrete, contradiction. So, we have finally proved (1).

Let Y be a subset of $\mathrm{Ult}A$ which is dense in $\mathrm{Ult}A$ and of cardinality $\mathrm{d}A$. Let

$$\mathcal{N} = \{\overline{Z} : Z \subseteq Y,\ |Z| \leq \mathrm{t}A\}.$$

From (1) and Lemma 11.5 we see that $|\mathcal{N}| \leq 2^{\mathrm{s}A}$. So, we will be finished, by Theorem 11.4, after we show that $\mathcal{N}$ is a network for A. Let $F \in U$, with U open. Say $F \in V \subseteq \overline{V} \subseteq U$, with V open. Choose $Z \subseteq Y$ with $|Z| \leq \mathrm{t}A$ such that $F \in \overline{Z}$. Let $Z' = V \cap Z$. Then $F \in \overline{Z'} \subseteq U$ and $\overline{Z'} \in \mathcal{N}$, as desired. □

By Theorem 11.1, spread can be considered to be an ordinary sup-function, and so its behaviour under unions is given by Theorem 1.1.

We turn to the derived functions for spread. The following facts are clear: $\mathrm{s}_{\mathrm{H}+}A = \mathrm{s}A$; $\mathrm{s}_{\mathrm{S}+}A = \mathrm{s}A$; $\mathrm{s}_{\mathrm{S}-}A = \omega$; $\mathrm{s}_{\mathrm{h}-}A = \omega$; ${}_{\mathrm{d}}\mathrm{s}_{\mathrm{S}+}A = \mathrm{s}A$. Again we have a problem for H-:

Problem 36. *Is* $\mathrm{s}_{\mathrm{H}-}A = \mathrm{Card}_{\mathrm{H}-}A$ *for every infinite BA A?*

It is also easy to see that $\mathrm{s}_{\mathrm{h}+}A = \mathrm{s}A$. The status of the derived function ${}_{\mathrm{d}}\mathrm{s}_{S-}$ is not clear; note that ${}_{\mathrm{d}}\mathrm{s}_{S-}A < \mathrm{s}A$ for $A = \mathcal{P}\kappa$.

Turning to the relationships of spread to our other functions, we first list out the things already proved: $\mathrm{c_{H+}}A = \mathrm{s}A$ by Theorem 1.3; $\mathrm{Depth}_{h+}A = \mathrm{s}A$ in Theorem 2.13; $\mathrm{t}A \leq \mathrm{s}A$ in Theorem 11.5; and $|A| \leq 2^{\mathrm{s}A}$ in Theorem 11.7. Note that $\mathrm{t}A$ can be much smaller than $\mathrm{s}A$, for example in the finite-cofinite algebra on an infinite cardinal κ. Also note that, obviously, $\mathrm{c}A \leq \mathrm{s}A$; and the difference is big in, e.g., free algebras. We have $\mathrm{s}A > \mathrm{Length}A$ for A a free algebra; $\mathrm{s}A < \mathrm{Length}A$ for A the interval algebra on the reals. Also, $\mathrm{s}A > \pi A$ for $A = \mathcal{P}\kappa$. The interval algebra of a Souslin line provides an example of a BA A with $\mathrm{s}A = \omega$ and $\mathrm{d}A > \omega$. In fact, clearly $\mathrm{s}A = \mathrm{c}A$ for A an interval algebra, by the retractiveness of interval algebras.

An example with $\mathrm{s}A = \omega < \mathrm{d}A$ cannot be given in ZFC; this follows from the following rather deep results. In Juhász [71] showed that under the assumption of MA+¬CH, for every compact Hausdorff space X, if $\mathrm{hL}X = \omega$ then $\mathrm{hd}X = \omega$. Todorčević [83] showed that it is consistent with MA+¬CH that for every regular space X, if $\mathrm{s}X = \omega$ then $\mathrm{hL}X = \omega$. Hence it is consistent that for every BA A, if $\mathrm{s}A = \omega$ then $\mathrm{hl}A = \omega = \mathrm{hd}A$.

12. CHARACTER

First note that we can define χA as a sup; namely, for any ultrafilter F on A let $\chi F = \min\{|X| : X \text{ is a set of generators of } F\}$—then $\chi A = \sup\{\chi F : F \text{ is an ultrafilter on } A\}$. Clearly then, by topological duality, $\chi(A \times B) = \sup(\chi A, \chi B)$. For a weak product we have $\chi(\prod^{\text{w}}_{i\in I} A_i) = \max(|I|, \sup_{i\in I}\chi A_i)$. To show this, it suffices to show that $\chi F = |I|$ for the "new" ultrafilter F. This ultrafilter is defined as follows. For each subset M of I, let x_M be the element of $\prod_{i\in I} A_i$ such that $x_M i = 1$ if $i \in M$ and $x_M i = 0$ for $i \notin M$. Then F is the set of all $y \in \prod^{\text{w}}_{i\in I} A_i$ such that $x_M \leq y$ for some cofinite subset M of I. So, it is clear that $\chi F \leq |I|$. If X is a set of generators for F with $|X| < |I|$, then there is a $y \in X$ such that $y \subseteq x_M$ for infinitely many cofinite subsets M of I; this is clearly impossible.

As usual, weak products enable us to discuss the attainment problem. Any infinite BA has a non-principal ultrafilter, and hence if A has character ω, then it is attained. Next, if κ is a singular cardinal, then we can construct a BA A with $\chi A = \kappa$ not attained. Namely, let $\langle \mu_\xi : \xi < \text{cf}\kappa\rangle$ be an increasing sequence of infinite cardinals with sup κ. For each $\xi < \text{cf}\kappa$ let A_ξ be the free BA on μ_ξ free generators; thus $\chi A_\xi = \mu_\xi$. By the above remarks on weak products, $\prod^{\text{w}}_{\xi<\text{cf}\kappa} A_\xi$ has character κ not attained. In the case of an uncountable regular limit cardinal κ, GCH implies that every BA with character κ has character attained. This follows from the theorem below that $|\text{Ult}A| \leq 2^{\chi A}$. Without GCH we do not know the answer:

Problem 37. *Can one prove without GCH that if χA is an uncountable regular limit cardinal then χA is attained?*

To treat arbitrary direct products, note that obviously $\text{t}A \leq \chi A$; hence $\text{Ind}A \leq \chi A$, and so clearly $\chi(\prod_{i\in I} A_i) \geq \max(2^{|I|}, \sup_{i\in I}\chi A_i)$. We do not know whether equality always holds:

Problem 38. *Is $\chi(\prod_{i\in I} A_i) = \max(2^{|I|}, \sup_{i\in I}\chi A_i)$?*

Character can increase in going from an algebra to a subalgebra. To construct an example of this sort, first notice that if A is the finite-cofinite algebra on an infinite cardinal κ, then $\chi A = \kappa$, by our initial remarks (since $A = \prod^{\text{w}}_{\alpha<\kappa} 2$). The algebra that we want is the Alexandroff duplicate of the free algebra on κ free generators, where κ is any infinite cardinal. Recall from page 19 the definition of the Alexandroff duplicate. Now let B be the free BA on κ free generators, κ any infinite cardinal. We claim that $\chi\text{Dup}B = \kappa$. To see this, we describe the ultrafilters on $\text{Dup}B$. Note that $\text{Dup}B$ is atomic, and its atoms are all of the elements $(0, \{F\})$ for $F \in \text{Ult}A$. So there is a principal ultrafilter corresponding to each of these atoms. Next, if G is an ultrafilter on B, then $G^+ \stackrel{\text{def}}{=} \{(a, X) : a \in G, X \subseteq \text{Ult}B, \mathcal{S}a \triangle X \text{ finite}\}$ is an ultrafilter on Dup$B$. Conversely, any nonprincipal ultrafilter on Dup$B$ is easily seen to have this form. Thus it suffices to show that any ultrafilter of this form has character κ. So, let F be an arbitrary ultrafilter on B. We claim that the set X of all elements of F^+ of the form $(a, \mathcal{S}a)$ generates F^+. For, let (a, Y) be any element of F^+; thus $\mathcal{S}a \backslash Y$ is finite. Clearly there is an element b of F such that $b \leq a$ and $b \notin G$ for every $G \in \mathcal{S}a \backslash Y$. Then $(b, \mathcal{S}b)$ is the desired member of X which is $\leq (a, Y)$. So, this shows that $\chi F^+ \leq \kappa$. An easy argument shows that actually $\chi F^+ = \kappa$. Namely, if Z generates F^+ and $|Z| < \kappa$, then choose $(a, X) \in Z$ such that $(a, X) \leq (b, \mathcal{S}b)$ for infinitely many $b \in F$ such that b or $-b$ is one of the free generators of B; this is impossible. So, $\chi\text{Dup}B = \kappa$. But the finite-cofinite algebra A on Ult$B$ is isomorphic to a subalgebra of Dup$B$, and by the previous remarks it has character 2^κ.

If A is a homomorphic image of B, then $\chi A \leq \chi B$ (let f be a homomorphism from B onto A; if $F \in \text{Ult}A$, then $f^{-1}[F] \in \text{Ult}B$, and if we choose $X \subseteq f^{-1}[F]$ with $|X| \leq \chi B$ such that X generates $f^{-1}[F]$, then $f[X]$ generates F). It is also easy to see that if $\langle A_i : i \in I \rangle$ is a system of BA's each with at least four elements, then $\chi(\bigoplus_{i \in I} A_i) = \max(|I|, \sup_{i \in I} \chi A_i)$. In fact, for $\geq$, first let $j \in I$ and let $F \in \text{Ult}A_j$. Let G be any ultrafilter on $\bigoplus_{i \in I} A_i$ which includes F. Suppose that $X \subseteq G$ generates G. Without loss of generality, each member of X is a product of elements from distinct A_i's. Then it is clear that $X \cap A_j \subseteq F$ and $X \cap A_j$ generates F. So $\chi F \leq |X|$. It follows that $\chi F \leq \chi G \leq \chi(\bigoplus_{i \in I} A_i)$. Hence $\chi A_j \leq \chi(\bigoplus_{i \in I} A_i)$. It is clear that $\chi H \geq |I|$ for any ultrafilter H on $\bigoplus_{i \in I} A_i$. Altogether, this proves $\geq$. For $\leq$, for any ultrafilter G on $\bigoplus_{i \in I} A_i$, and for each $i \in I$ let $X_i \subseteq G \cap A_i$ generate $G \cap A_i$, with $|X_i| = \chi(G \cap A_i)$. Clearly the set of all finite products of elements of $\bigcup_{i \in I} X_i$ generates G, as desired.

We have not investigated how character acts under the union of well-ordered chains.

We turn to the derived functions for character. By a remark above, we have $\chi_{\text{H}+} A = \chi A$ for any infinite BA A. We do not know about $\chi_{\text{H}-}$; in particular, the following problem is open.

Problem 39. *Is* $\chi_{\text{H}-} A = \text{Card}_{\text{H}-} A$ *for any infinite BA* A?

We also do not know the status of $\chi_{\text{S}+} A$; we observed above that it can happen that $\chi_{\text{S}+} A > \chi A$. Clearly $\chi_{\text{S}-} A = \omega$ for any infinite BA A. The topological version of character is this: for any space X and any $x \in X$, $\chi(x, X)$ is the minimum of the cardinalities of neighborhood bases for x in X, and $\chi X = \sup\{\chi(x, X) : x \in X\}$. Clearly then $\chi_{\text{h}+} A = \chi A$, and $\chi_{\text{h}-} A = 1$ for any infinite BA A, since A has an infinite discrete subspace. We do not know the status of the derived functions ${}_d\chi_{\text{S}+}$ and ${}_d\chi_{\text{S}-}$.

The function χ_{inf} is of some interest; recall from the introduction that

$\chi_{\inf}A = \inf\{\chi F : F \in \text{Ult}A\}$ for any infinite BA A. It has not been investigated much, but we give the following classical result of Čech and Pospíšil concerning it:

Theorem 12.1. *$2^{\chi_{\inf}A} \leq |\text{Ult}A|$ for any infinite BA A.*

PROOF. For brevity set $\kappa = \chi_{\inf}A$. It clearly suffices to construct a function f mapping ${}^{<\kappa}2$ into A such that

(1) For each $s \in {}^{<\kappa} 2$, the set $\{f(s \restriction \alpha) : \alpha \leq \text{dom}s\}$ has the finite intersection property;

(2) $f(s^\frown 0) \cdot f(s^\frown 1) = 0$ for each $s \in {}^{<\kappa} 2$.

Suppose $s \in {}^{<\kappa} 2$ and $f(s \restriction \alpha)$ has been defined for all $\alpha \in \text{dom}s$. By the induction hypothesis, $\{f(s \restriction \alpha) : \alpha \in \text{dom}s\}$ has the finite intersection property; since this set has $< \kappa$ elements, it does not generate an ultrafilter, and hence there is a $a \in A$ such that both a and $-a$ fail to be in the filter generated by it. Hence if we set $f(s^\frown 0) = a$ and $f(s^\frown 1) = -a$ we extend our function f so that (1) and (2) will hold. This completes the proof. □

We turn to the relationship of character with our previously treated functions. Obviously $\text{t}A \leq \chi A$ for any infinite BA A; the difference can be big—for example for the finite-cofinite algebra on an infinite cardinal κ. Now consider the possibility that $\chi A > \text{s}A$. By the comment at the end of the last section, plus the fact that $\chi A \leq \text{hL}A$ (easy, and proved in section 13), it is consistent that $\text{s}A = \omega$ implies $\chi A = \omega$. The Kunen line, constructed in section 6, has uncountable character but countable spread; it was constructed using CH.

Problem 40. *Can one construct in ZFC a BA A such that $\text{s}A < \chi A$?*

An example of a BA A with $\text{c}A > \chi A$ is provided by the Alexandroff duplicate of the free algebra on κ free generators, as discussed above. The interval algebra on **R** gives an example of an algebra A with $\text{Length}A > \chi A$.

Now we turn to Arhangelskiĭ's theorem that $|\text{Ult}A| \leq 2^{\chi A}$ for any infinite BA A. We need three lemmas which are also of interest. For any set W of ultrafilters we set $\overline{W} = \{K \in \text{Ult}A : K \subseteq \bigcup W\}$; this coincides with the topological closure of W.

Lemma 12.2. *If $Y \subseteq \text{Ult}A$ and $|Y| \leq \chi A$, then $|\overline{Y}| \leq 2^{\chi A}$.*

PROOF. For every ultrafilter G on A let $\{a^G_\alpha : \alpha < \chi A\}$ be a set of generators of G, and set $fG = \{\{F \in Y : a^G_\alpha \in F\} : \alpha < \chi A\}$. Thus $fG \subseteq \mathcal{P}Y$. Hence it is enough to show that $f \restriction \overline{Y}$ is one-one. Suppose that G and H are distinct ultrafilters on A such that $G, H \in \overline{Y}$. Say $a^G_\alpha \in G \backslash H$, and choose $a^H_\beta \leq -a^G_\alpha$. Suppose that $fG = fH$; then there is a $\gamma < \chi A$ such that $\{F \in Y : a^G_\gamma \in F\} = \{F \in Y : a^H_\beta \in F\}$. Then $a^G_\alpha \cdot a^G_\gamma \in G$; say then $a^G_\alpha \cdot a^G_\gamma \in F \in Y$. Then $a^H_\beta \in F$, $a^G_\alpha \in F$, and $-a^G_\alpha \in F$, contradiction. □

Lemma 12.3. *If $Y \subseteq \mathrm{Ult}A$ and $F \subseteq \bigcup Y$ is closed under $\cdot$ and has the finite intersection property, then there is an ultrafilter G with the property that $F \subseteq G \subseteq \bigcup Y$.*

PROOF. Let G be maximal among the filters H such that $F \subseteq H \subseteq \bigcup Y$. Suppose that G is not an ultrafilter; say $a, -a \notin G$. Then $\langle G \cup \{a\}\rangle^{\mathrm{Fi}} \not\subseteq \bigcup Y$. Say $b \in G$ and $b \cdot a \notin \bigcup Y$. Similarly obtain $c \in G$ such that $c \cdot -a \notin \bigcup Y$. Choose $H \in Y$ such that $b \cdot c \in H$. Then $b \cdot c \cdot a \notin H$ and $b \cdot c \cdot -a \notin H$, contradiction. □

A subset $Y \subseteq \mathrm{Ult}A$ is *closed* iff $Y = \{F \in \mathrm{Ult}A : F \subseteq \bigcup Y\}$. This coincides with the topological notion.

Lemma 12.4. *If $Z \subseteq \mathrm{Ult}A$ is closed and $|Z| \leq 2^{\chi A}$, then $\mathrm{Ult}A \backslash Z$ is a union of $\leq 2^{\chi A}$ closed-open sets.*

PROOF. For every $G \in \mathrm{Ult}A$ let $\{a^G_\alpha : \alpha < \chi A\}$ be a set of generators of G, closed under $\cdot$. Let

$$B = \{a^{G_0}_{\alpha_0} + \cdots + a^{G_{n-1}}_{\alpha_{n-1}} : n \in \omega,\ G_0, \ldots, G_{n-1} \in Z,\ \alpha_0, \ldots, \alpha_{n-1} < \chi A\},$$

and let $C = \{y : -y \in B \cap \bigcap Z\}$. We claim that

$$\mathrm{Ult}A \backslash Z = \bigcup_{y \in C} \mathcal{S}y,$$

which gives the desired result. $\supseteq$ is clear. Now suppose that $F \in \mathrm{Ult}A \backslash Z$. For every $G \in Z$ choose $b_G \in F \backslash G$; say $a^G_{\alpha(G)} \leq -b_G$. Then

(*) There exist an integer $n \in \omega$ and elements $G_0, \ldots, G_{n-1} \in Z$ with the property that $a^{G_0}_{\alpha(G_0)} + \cdots + a^{G_{n-1}}_{\alpha(G_{n-1})} \in H$ for all $H \in Z$.

Otherwise, $L \stackrel{\text{def}}{=} \{-a^{G_0}_{\alpha(G_0)} \cdot \cdots \cdot -a^{G_{n-1}}_{\alpha(G_{n-1})} : n \in \omega,\ G_0, \ldots, G_{n-1} \in Z\}$ has the finite intersection property, is closed under $\cdot$, and is contained in $\bigcup Z$. Hence by Lemma 12.3, there is an ultrafilter K such that $L \subseteq K \subseteq \bigcup Z$. Hence $K \in Z$ and $-a^K_{\alpha(K)} \in K$, contradiction.

We choose $n \in \omega$ and $G_0, \ldots, G_{n-1} \in Z$ as in (*). Let y be the element $-a^{G_0}_{\alpha(G_0)} \cdot \cdots \cdot -a^{G_{n-1}}_{\alpha(G_{n-1})}$. Then $y \in C$ and $F \in \mathcal{S}y$, as desired. □

Now we are ready for the proof of Arhangelskiĭ's theorem:

Theorem 12.5. $\mathrm{Ult}A \leq 2^{\chi A}$ *for any infinite BA A.*

PROOF. Suppose that $2^{\chi A} < |\mathrm{Ult}A|$. Fix an ultrafilter F on A. For each $f \in \bigcup_{\alpha<(\chi A)^+} {}^{\alpha}(2^{\chi A})$ we define a closed set $X_f \subseteq \mathrm{Ult}A$ and a $G_f \in \mathrm{Ult}A$. Let $X_0 = \mathrm{Ult}A$ and $G_0 = F$. For $\mathrm{dom}f$ limit let $X_f = \bigcap_{\alpha<\mathrm{dom}f} X_{f\restriction\alpha}$, and if $X_f \neq 0$ choose $G_f \in X_f$, and otherwise let $G_f = F$. Now suppose that $\mathrm{dom}f$ is a successor ordinal $\alpha + 1$. Let $g = f \restriction \alpha$, and set $Y_g = \{G_{g\restriction\beta} : \beta \leq \alpha\}$. Thus $|Y_g| \leq \chi A$, so by Lemma 12.2, $|\overline{Y}_g| \leq 2^{\chi A}$, and so by Lemma 12.4 we can let $\langle a^g_\beta : \beta < 2^{\chi A}\rangle$ be such that $\mathrm{Ult}A\backslash\overline{Y}_g = \bigcup_{\beta<2^{\chi A}} \mathcal{S}(a^g_\beta)$ and set $X_f = X_g \cap \mathcal{S}(a^g_{f\alpha})$. Again let $G_f \in X_f$ if $X_f \neq 0$, and $G_f = F$ otherwise. This finishes the construction.

Now choose

$$H \in \mathrm{Ult}A\backslash\bigcup\{\overline{\{G_{f\restriction\beta} : \beta \leq \mathrm{dom}f\}} : f \in \bigcup_{\alpha<(\chi A)^+} {}^{\alpha}(2^{\chi A})\}$$

Now we define f mapping $(\chi A)^+$ into $2^{\chi A}$ by induction. Suppose that $f\beta$ has been defined for all $\beta < \alpha$. Now $H \notin \overline{\{G_{f\restriction\beta} : \beta \leq \alpha\}}$, so there is a $\gamma < 2^{\chi A}$ such that $H \in X_{(f\restriction\alpha)\cup\{(\alpha,\gamma)\}}$; set $f\alpha = \gamma$. Thus $H \in X_{f\restriction\alpha}$ for all $\alpha < (\chi A)^+$. We claim that $\langle G_{f\restriction(\beta+1)} : \beta < (\chi A)^+\rangle$ is a free sequence, which contradicts $\mathrm{t}A \leq \chi A$. Let $\alpha < (\chi A)^+$ and suppose that

$$K \in \overline{\{G_{f\restriction(\beta+1)} : \beta < \alpha\}} \cap \overline{\{G_{f\restriction(\beta+1)} : \alpha \leq \beta < (\chi A)^+\}}$$

Then $K \in \overline{\{G_{f\restriction\beta} : \beta \leq \alpha\}}$, so $K \in \mathrm{Ult}A\backslash X_{f\restriction(\alpha+1)}$, and this set is open, so there is a $\beta \geq \alpha$ such that $G_{f\restriction(\beta+1)} \in \mathrm{Ult}A\backslash X_{f\restriction(\alpha+1)} \subseteq \mathrm{Ult}A\backslash X_{f\restriction(\beta+1)}$, contradiction. □

We describe character for interval algebras. Let L be an ordering, and A the interval algebra on L. As mentioned at the end of section 9, the ultrafilters on A are in one-one correspondence with the terminal segments T of L. The *character* of such a terminal segement is the pair (κ, λ^*) such that $L\backslash T$ has cofinality κ and T has coinitiality λ. And χF is the maximum of κ and λ. χA is the supremum of all χF. From this description it is clear that $\mathrm{Depth}A = \chi A$ (and hence both are equal to $\mathrm{t}A$), for any interval algebra A.

13. HEREDITARY LINDELÖF DEGREE

We begin with some equivalent definitions. For one of them we need the following notion: a sequence $\langle x_\xi : \xi < \kappa\rangle$ of distinct elements of a topological space X is *right-separated* provided that for every $\xi < \kappa$ the set $\{x_\eta : \eta < \xi\}$ is open in $\{x_\xi : \xi < \kappa\}$.

Theorem 13.1. *For any infinite BA A,* hLA *is equal to each of the following cardinals:*
sup$\{\kappa$:*there is an ideal not generated by less than* κ *elements*$\}$;
sup$\{\kappa$:*there is a strictly increasing sequence of ideals of length* $\kappa\}$;
sup$\{\kappa$:*there is a strictly increasing sequence of filters of length* $\kappa\}$;
sup$\{\kappa$:*there is a strictly increasing sequence of open sets of length* $\kappa\}$;
sup$\{\kappa$:*there is a strictly decreasing sequence of closed sets of length* $\kappa\}$;
sup$\{\kappa$:*there is a right-separated sequence of length* $\kappa\}$;
min$\{\kappa$:*every open cover of a subspace of* Ult$A$ *has a subcover of size* $\leq \kappa\}$.

PROOF. Eight cardinals are mentioned; let them be denoted by $\kappa_0, \ldots, \kappa_7$ in their order of mention (starting with hL). First we take care of easy relations: $\kappa_2 = \kappa_3$ since, if I is an ideal then $I^f \stackrel{\text{def}}{=} \{a \in A : -a \in I\}$ is a filter, and $I \subset J$ iff $I^f \subset J^f$; similarly, going from filters to ideals. So $\kappa_2 = \kappa_3$ follows. Next, $\kappa_2 \leq \kappa_4$. For, if I is an ideal, let $I^u = \bigcup_{a \in I} \mathcal{S}a$. Then I^u is open, and $I \subset J$ implies $I^u \subset J^u$. (If $a \in J \backslash I$, then $\mathcal{S}a \subseteq J^u$, of course, but $\mathcal{S}a \not\subseteq I^u$, since otherwise compactness of $\mathcal{S}a$ would easily yield $a \in I$.) This shows $\kappa_2 \leq \kappa_4$. It is clear that $\kappa_4 = \kappa_5$, by taking complements. $\kappa_4 \leq \kappa_6$: If $\langle U_\alpha : \alpha < \kappa\rangle$ is a strictly increasing sequence of open sets, for every $\alpha < \kappa$ choose $x_\alpha \in U_{\alpha+1} \backslash U_\alpha$. Clearly $\langle x_\alpha : \alpha < \kappa\rangle$ is right-separated. $\kappa_7 \leq \kappa_0$: for any subspace X of Ult$A$, any cover of X has a subcover of power $\leq \mathrm{L}X \leq \kappa_0$, so $\kappa_7 \leq \kappa_0$.

It remains only to prove that $\kappa_0 \leq \kappa_1$, $\kappa_1 \leq \kappa_2$, and $\kappa_6 \leq \kappa_7$. For the first one, suppose that $X \subseteq$ Ult$A$ and $\mathcal{O}$ is an open cover of X with no subcover of power λ; we construct an ideal not generated by λ or fewer elements. Let

$$I = \langle\{a \in A : \exists U \in \mathcal{O}(\mathcal{S}a \cap X \subseteq U)\}\rangle^{\mathrm{Id}}.$$

Suppose that I is generated by J, where $|J| \leq \lambda$. For every $a \in J$ there is a finite subset $\mathcal{P}_a$ of $\mathcal{O}$ such that $\mathcal{S}a \cap X \subseteq \bigcup \mathcal{P}_a$. Let $\mathcal{O}' = \bigcup_{a \in J} \mathcal{P}_a$. We claim that $\mathcal{O}'$ covers X, which is the desired contradiction. Indeed, let $x \in X$. Say $x \in U \in \mathcal{O}$. Say $x \in \mathcal{S}a \cap X \subseteq U$. Choose a finite subset F of J such that $a \leq \sum F$. Then $x \in \mathcal{S}a \cap X \subseteq \bigcup_{a \in F} \mathcal{P}a$, as desired.

Next, $\kappa_1 \leq \kappa_2$: suppose that I is an ideal not generated by fewer than λ elements. Then it is easy to construct a sequence $\langle a_\alpha : \alpha < \lambda\rangle$ of elements of I such that $a_\alpha \notin \langle\{a_\beta : \beta < \alpha\rangle^{\mathrm{Id}}$ for all $\alpha < \lambda$. Thus $\langle\langle\{a_\beta : \beta < \alpha\rangle^{\mathrm{Id}} : \alpha < \lambda\rangle$ is a strictly increasing sequence of ideals, as desired.

For $\kappa_6 \leq \kappa_7$, suppose that λ is a regular cardinal $\leq \kappa_5$ and $\langle x_\alpha : \alpha < \lambda \rangle$ is right separated. Thus for each $\alpha < \lambda$ we can choose an open set U_α such that $U_\alpha \cap \{x_\xi : \xi < \lambda\} = \{x_\xi : \xi < \alpha + 1\}$. Then $\{U_\alpha : \alpha < \lambda\}$ is a cover of $\{x_\xi : \xi < \lambda\}$ with no subcover of size $< \lambda$. Hence $\lambda < \kappa_7$, and this shows that $\kappa_6 \leq \kappa_7$. □

In Theorem 13.1, seven of the eight equivalents involve sups, and thus give rise to attainment problems. The proof of the theorem shows the following: attainment is the same for κ_2 and κ_3, and for κ_4 and κ_5; moreover, attainment in the sense κ_2 implies attainment in the sense κ_4, attainment in the sense κ_4 implies attainment in the sense κ_6, and attainment in the sense κ_1 implies attainment in the sense κ_2. It is also easy to see that attainment in the sense κ_4 implies attainment in the sense κ_2. In fact, if $\langle U_\alpha : \alpha < \kappa_3 \rangle$ is an increasing sequence of open sets, for each $\alpha < \kappa_3$ let $I_\alpha = \{a : \mathcal{S}a \subseteq U_\alpha\}$. Clearly I_α is an ideal. To show properness, pick $F \in U_{\alpha+1} \backslash U_\alpha$. Say $F \in \mathcal{S}a \subseteq U_{\alpha+1}$. Thus $a \in I_{\alpha+1} \backslash I_\alpha$. And attainment in the sense κ_6 implies attainment in the sense κ_4. In fact, suppose that $\langle F_\alpha : \alpha < \kappa \rangle$ is right separated. For each $\alpha < \kappa$ choose $a_\alpha \in F_\alpha$ such that $\mathcal{S}a_\alpha \cap \{F_\beta : \beta > \alpha\} = 0$, and let $U_\alpha = \bigcup_{\beta<\alpha} \mathcal{S}a_\beta$. Note that $F_\alpha \in U_{\alpha+1} \backslash U_\alpha$, as desired.

Thus we have seen that there are only three versions of the definition of hL that might lead to different attainment properties. We do not know whether, in fact, they are different:

Problem 41. *Does attainment of* hL *imply attainment in the right-separated sense?*

Problem 42. *Does attainment of* hL *in the right-separated sense imply attainment in the defined sense?*

Problem 43. *Does attainment of* hL *in the right separated sense imply attainment in the sense of generation of ideals?*

It is known that hL is attained in the right-separated sense for cardinals of cofinality ω, and for strong limit singular cardinals; see Juhász [80].

We turn to algebraic operations. If A is a subalgebra or homomorphic image of B, then $\mathrm{hL}A \leq \mathrm{hL}B$. Furthermore, looking at the right-separated equivalent and the topological dual it is clear that

$$\mathrm{hL}\Big(\prod_{i\in I}^{\mathrm{w}} A_i\Big) = \max(|I|, \sup_{i\in I}\mathrm{hL}A).$$

Note that $\mathrm{Ind}A \leq \mathrm{hL}A$, using the equivalent concerning ideals, for example. Hence it is clear that $\mathrm{hL}(\prod_{i\in I} A_i) \geq \max(2^{|I|}, \sup_{i\in I}\mathrm{hL}A_i)$, but we do not know whether equality holds:

Problem 44. *Is* $\mathrm{hL}(\prod_{i\in I} A_i) = \max(2^{|I|}, \sup_{i\in I}\mathrm{hL}A_i)$ *for non-trivial BA's* A_i*?*

Next come free products:

Theorem 13.2. *If* $\langle A_i : i \in I\rangle$ *is a system of non-trivial BA's, for brevity let* $\lambda = \sup_{i\in I}\mathrm{hL}A_i$*; then*

$$\max(|I|, \sup_{i\in I}\mathrm{hL}A_i) \leq \mathrm{hL}(\oplus_{i\in I}A_i) \leq \max(|I|, 2^\lambda).$$

PROOF. The first inequality follows immediately from the preceeding remarks. We consider two cases.

Case 1. $|I| \leq \lambda$. Suppose $\mathrm{hL}(\oplus_{i\in I}A_i) > 2^\lambda$; we shall get a contradiction. Let $\langle F_\alpha : \alpha < (2^\lambda)^+\rangle$ be right-separated in Ult$A$. For each $\alpha < \lambda$ choose $a_\alpha \in F_\alpha$ such that for all $\beta < \lambda$, $a_\alpha \in F_\beta$ implies $\beta \leq \alpha$. We may assume that a_α has the form $\prod G_\alpha$, where G_α is a finite set of elements of $\bigcup_{i\in I} A_i$. If $\beta < \alpha$, choose $b_{\alpha\beta} \in G_\beta$ such that $b_{\alpha\beta} \notin F_\alpha$; say $b_{\alpha\beta} \in A_{k\alpha\beta}$. By $(2^\lambda)^+ \to (\lambda^+)_\lambda$, there is an $M \in [(2^\lambda)^+]^{\lambda^+}$ and a $k \in I$ such that for all $\beta, \alpha \in M$ with $\beta < \alpha$ we have $k\alpha\beta = k$. Then if $\beta \in M$, we have $b_{\alpha\beta} \in F_\beta \cap A_k$ and for all $\alpha \in M$ with $\beta < \alpha$ we have $b_{\alpha\beta} \notin F_\alpha \cap A_k$, so $\langle F_\beta \cap A_k : \beta \in M\rangle$ is right separated in A_k, contradiction.

Case 2. $|I| > \lambda$. Suppose that $\langle F_\alpha : \alpha < (\max(|I|, 2^\lambda))^+\rangle$ is right-separated; choose a_α and G_α as above, and say that $G_\alpha \subseteq \bigcup_{i\in J_\alpha} A_i$, where J_α is a finite subset of I. We may assume that $J_\alpha = J$ for all α; but then Case 1 is contradicted. □

Concerning derived functions of hL, we mention these obvious facts:

$$\mathrm{hL}A = \mathrm{hL}_{\mathrm{H}+}A = \mathrm{hL}_{\mathrm{S}+}A = \mathrm{hL}_{\mathrm{h}+}A = {}_{\mathrm{d}}\mathrm{hL}_{\mathrm{S}+}A;$$

and $\mathrm{hL}_{\mathrm{S}-}A = \mathrm{hL}_{\mathrm{h}-}A = \omega$. We do not know the status of $\mathrm{hL}_{\mathrm{H}-}$ and ${}_{\mathrm{d}}\mathrm{hL}_{\mathrm{S}-}$.

On the relationship of hL with the previously defined functions: obviously $\mathrm{s}A \leq \mathrm{hL}A$ for any infinite BA A. Next, $\chi A \leq \mathrm{hL}A$. In fact, suppose that F is any ultrafilter on A; we want to find a subset X of F which generates F and has at most $\mathrm{hL}A$ elements. The set $\{\mathcal{S}a : -a \in F\}$ covers Ult$A\backslash\{F\}$. Hence there is a subset X of F such that $\{\mathcal{S}a : -a \in X\}$ also covers Ult$A\backslash\{F\}$, and $|X| \leq \mathrm{hL}A$. We claim that X generates F. For suppose that $a \in F$. Then $X \cup \{-a\}$ does not have the finite intersection property; otherwise, there would exist an ultrafilter G containing this set—then $G \neq F$, so $b \in G$ for some b such that $-b \in X$, contradiction. But $X \cup \{-a\}$ not having the finite intersection property means that a is in the filter generated by X, as desired.

The BA on the Kunen line constructed in section 6 (assuming CH) has character ω_1, hence hereditary Lindelöf degree ω_1, and countable spread. To show that its character is ω_1, assume the notation of section 6. We show that y, the "new" point in the one-point compactification, has character ω_1. Suppose that y has a countable base $\mathcal{V}$. We may assume that each member of $\mathcal{V}$ has the form $\{y\} \cup (\mathbf{R}\backslash V)$, where $V \subseteq \mathbf{R}$ is compact open in τ (see the end of the discussion of Example 6.1, just preceding (12)). Now for such a V we have $V = \bigcup_{\alpha<\omega_1} V \cap \mathbf{R}_\alpha$, so compactness of V yields that V is countable. But y has neighborhoods of the form $\{y\} \cup (\mathbf{R}\backslash(\{x_\alpha\} \cup V))$ for each $\alpha < \omega_1$, so this clearly leads to a contradiction.

An example where $\chi A < \mathrm{hL}A$ is provided by the Alexandroff duplicate of a free algebra; see section 12. An example with $\mathrm{hL}A < \mathrm{d}A$ is provided by the interval algebra on a complete Souslin line, using the argument of Lemma 1.6.; on the other hand, in the first edition of Juhász's book, it is shown that MA+¬CH implies that $\mathrm{hL}A = \omega$ implies $\mathrm{hd}A = \omega$.

These observations leave the following two questions open.

Problem 45. *Is there an example in ZFC of a BA A such that* $\mathrm{s}A < \mathrm{hL}A$*?*

Problem 46. *Is there an example in ZFC of a BA A such that* $\mathrm{hL}A < \mathrm{d}A$*?*

For an interval algebra A we have $\mathrm{hL}A = \mathrm{c}A$. In fact, suppose that A is the interval algebra on L but $\mathrm{hL}A > \mathrm{c}A$. Let $\langle I_\alpha : \alpha < \kappa \rangle$ be a strictly increasing sequence of ideals in A, where $\kappa = (\mathrm{c}A)^+$. Define $a \equiv_\alpha b$ iff either $a = b$ or else if, say, $a < b$, then $[a, b) \in I_\alpha$. Then $\equiv_\alpha$ is a convex equivalence relation on L, and hence has $\leq \mathrm{c}A$ classes. The left endpoints of the $\equiv_\alpha$ classes (in the completion of L) are decreasing, and the right endpoints, increasing. Since the completion of L cannot have a strictly increasing or strictly decreasing sequence of type $(\mathrm{c}A)^+$, it follows that for each $\equiv_\kappa$ class, there is an $\alpha < \kappa$ such that the given class is constant from the αth stage on, which is clearly impossible.

14. HEREDITARY DENSITY

We begin again with some equivalent definitions, which are similar to the case of hereditary Lindelöf degree. Recall from page 42 the definition of left-separated sequence.

Theorem 14.1. *For any infinite BA A,* hd$A$ *is equal to each of the following cardinals:*

sup$\{\kappa$:*there is a strictly decreasing sequence of ideals of length* $\kappa\}$;
sup$\{\kappa$:*there is a strictly decreasing sequence of filters of length* $\kappa\}$;
sup$\{\kappa$:*there is a strictly decreasing sequence of open sets of length* $\kappa\}$;
sup$\{\kappa$:*there is a strictly increasing sequence of closed sets of length* $\kappa\}$;
sup$\{\kappa$:*there is a left-separated sequence of length* $\kappa\}$;
min$\{\kappa$:*every subspace S of* Ult$A$ *has a dense subset of power* $\leq \kappa\}$;
sup$\{\pi B : B$ *is a homomorphic image of* $A\}$;
sup$\{\mathrm{d}B : B$ *is a homomorphic image of* $A\}$.

PROOF. This time there are nine cardinals, named $\kappa_0, \ldots, \kappa_8$ in their order of mention. The following relationships are easy, following the pattern of the proof of Theorem 13.1: $\kappa_1 = \kappa_2$; $\kappa_1 \leq \kappa_3$; $\kappa_3 = \kappa_4$; $\kappa_3 \leq \kappa_5$; and $\kappa_0 = \kappa_6$. Furthermore, $\kappa_8 = \kappa_0$ by Theorem 3.6, and $\kappa_0 = \kappa_5$ by Theorem 4.2, and $\kappa_0 = \kappa_7$ by Theorem 4.4. Hence only two inequalities remain.

$\kappa_6 \leq \kappa_2$: Suppose that X is a subspace of Ult$A$, and d$X = \kappa$; we construct a strictly decreasing sequence of filters of type κ. By induction let

$$F_\alpha \in X \backslash \overline{\{F_\beta : \beta < \alpha\}}$$

for each $\alpha < \kappa_6$. Then set $C_\alpha = \bigcap_{\beta \leq \alpha} F_\beta$. Thus $\langle C_\alpha : \alpha < \kappa_6 \rangle$ is a decreasing sequence of filters. It is strictly decreasing, since if $\alpha < \kappa_6$ we can choose $a \in F_{\alpha+1}$ such that $\mathcal{S}a \cap \{F_\beta : \beta \leq \alpha\} = 0$, so that $-a \in C_\alpha \backslash C_{\alpha+1}$.

$\kappa_5 \leq \kappa_6$: Suppose $\langle x_\alpha : \alpha < \kappa \rangle$ is left separated, where κ is regular. Clearly then $\{x_\alpha : \alpha < \kappa\}$ has no dense subset of power $< \kappa$. □

The equivalents in Theorem 14.1 give rise to eight possible attainment problems, on the face of it. The proofs show that the attainment is the same for $\kappa_1, \kappa_2, \kappa_3, \kappa_4$, and also for κ_0, κ_8; and attainment of hd implies attainment in the κ_7 sense, attainment in the κ_7 sense implies attainment in the κ_5 sense, and attainment in the κ_3 sense implies attainment in the κ_5 sense. Moreover, attainment in the κ_5 sense implies attainment in the κ_3 sense. In fact, suppose that $\langle F_\alpha : \alpha < \kappa \rangle$ is a left-separated sequence in Ult$A$. For each $\alpha < \kappa$ choose $a_\alpha \in F_\alpha$ such that $\mathcal{S}a_\alpha \cap \{F_\beta : \beta < \alpha\} = 0$. For each $\alpha < \kappa$ let $U_\alpha = \bigcup_{\beta \geq \alpha} \mathcal{S}a_\beta$. This is a decreasing sequence of open sets. It is strictly decreasing, since $F_\alpha \in U_\alpha \backslash U_{\alpha+1}$.

Finally, by the remark following the proof of Theorem 4.4, attainment in the κ_7 sense implies attainment in the κ_5 sense, and attainment in the κ_0 sense implies attainment in the κ_7 sense. So there are only two remaining problems concerning relationships of attainments:

Problem 47. *Does attainment of* hd *imply attainment in the* $\pi_{\mathrm{H}+}$ *sense?*

Problem 48. *Does attainment of* $\pi_{\mathrm{H}+}$ *imply attainment of* hd *in the left-separated sense?*

Like for hL, it is known that hd in the sense of left-separation is attained for singular cardinals of cofinality ω and for strong limit singular cardinals.

If A is a subalgebra or homomorphic image of B, then $\mathrm{hd}A \leq \mathrm{hd}B$. It is also clear that

$$\mathrm{hd}\left(\prod_{i\in I}^{\mathrm{w}} A_i\right) = \max(|I|, \sup_{i\in I}\mathrm{hd}A_i).$$

Obviously $\mathrm{s}A \leq \mathrm{hd}A$, and hence $\mathrm{Ind}A \leq \mathrm{hd}A$. It follows that for arbitrary products we have, as usual , $\mathrm{hd}(\prod_{i\in I} A_i) \geq \max(2^{|I|}, \sup_{i\in I}\mathrm{hd}A_i)$, but we do not know about equality:

Problem 49. $\mathrm{hd}(\prod_{i\in I} A_i) = \max(2^{|I|}, \sup_{i\in I}\mathrm{hd}A_i)$?

For free products, the analog of Theorem 13.2 holds, with essentially the same proof:

Theorem 14.2. *If* $\langle A_i : i \in I\rangle$ *is a system of non-trivial BA's, for brevity let* $\lambda = \sup_{i\in I}\mathrm{hd}A_i$*; then*

$$\max(|I|, \sup_{i\in I}\mathrm{hd}A_i) \leq \mathrm{hd}(\oplus_{i\in I}A_i) \leq \max(|I|, 2^\lambda). \qquad \square$$

Concerning derived functions, we have the following obvious facts:

$$\mathrm{hd}A = \mathrm{hd}_{\mathrm{H}+}A = \mathrm{hd}_{\mathrm{S}+}A = \mathrm{hd}_{\mathrm{h}+}A = {}_{\mathrm{d}}\mathrm{hd}_{\mathrm{S}+}A;$$

and $\mathrm{hd}_{\mathrm{S}-}A = \mathrm{hd}_{\mathrm{h}-}A = \omega$. Again, we do not know the status of $\mathrm{hd}_{\mathrm{H}-}$ and ${}_{\mathrm{d}}\mathrm{hd}_{\mathrm{S}-}$.

On the relationship of hd with the other functions, note also that by Theorem 14.1 we have $\pi A \leq \mathrm{hd}A$. πA is strictly less than $\mathrm{hd}A$ in $\mathcal{P}\kappa$, for example. And we have $\mathrm{s}A < \mathrm{hd}A$ for A the interval algebra on a Souslin line, and $\mathrm{hd}A < \chi A$ for a Kunen line (section 6). The following problems are open.

Problem 50. *Can one construct in ZFC a BA A such that* $\mathrm{s}A < \mathrm{hd}A$*?*

Problem 51. *Can one construct in ZFC a BA A such that* $\mathrm{hd}A < \chi A$*?*

15. INCOMPARABILITY

We begin with one important equivalent definition:

Theorem 15.1. *For any infinite BA A we have* $\mathrm{Inc}A=\sup\{|T| : T$ *is a tree included in* $A\}$.

(Note that when we say that T is a tree included in A, we mean merely that T is a subset of A which is a tree under the induced ordering; there is no assumption that incomparable elements (in T) are disjoint (in the dual of A).)

PROOF. Since any incomparable set is a tree having only roots, the inequality $\leq$ is clear. To show equality, suppose that κ is regular and A has no incomparable set of size κ; we show that A has no tree of size κ. Suppose T is a tree of size κ. By Theorem 4.25 of Part I of the BA handbook, A has a dense subset D of size $< \kappa$. Now each level of T is an incomparable set, and hence has fewer than κ elements. Hence T has at least κ levels. Let T' be a subset of T of power κ consisting exclusively of elements of successor levels. For each $d \in D$ let

$$M_d = \{t \in T : \text{if } s \text{ is the immediate predecessor of } t, \text{ then } d \leq t \cdot -s\}.$$

Thus $T' = \bigcup_{d\in D} M_d$, so there is a $d \in D$ such that $|M_d| = \kappa$. But then M_d is incomparable, contradiction: if $y, z \in M_d$ and $y < z$, then $y \leq u$ where u is the immediate predecessor of z, and $d \leq z \cdot -u$, hence $d \cdot y = 0$, contradicting $d \leq y$. □

Note that if Inc$A$ is attained, then it is obviously attained in the tree sense. The converse is not clear.

Problem 52. *If* Inc$A$ *is attained in the tree sense, is it also attained in the defined sense?*

Concerning attainment of Inc, several things are known. Milner and Pouzet [86] proved a general result, of which a special case is that if $\mathrm{Inc}A = \lambda$ with $\mathrm{cf}\lambda = \omega$, then Inc$A$ is attained. Shelah [83] has shown that it is consistent that for every singular λ with $\mathrm{cf}\lambda > \omega$ there is a BA A with incomparablity λ not attained. Todorčević has shown that if 2^ω is weakly inaccessible, then there is a BA of size 2^ω with incomparability 2^ω not attained. On the other hand, Theorem 4.25 of the BA handbook shows that if Inc$A$ is a strong limit cardinal, then Inc$A$ is attained.

Now we turn to algebraic operations, as usual. If A is a subalgebra or homomorphic image of B, then $\mathrm{Inc}A \leq \mathrm{Inc}B$. If A is a subalgebra of B, then, easily, $\mathrm{Inc}(A \times B) \geq |A|$; in fact, $\{(a, -a))\}$ is an incomparable set in $A \times B$. Hence if A is cardinality-homogeneous and has no incomparable set of size $|A|$, then A is rigid (this follows from some elementary facts concerning automorphisms; see the article in the BA handbook about automorphisms). Thus the incomparability of a product can jump from that in a factor—for example, if A is such that $\mathrm{Inc}A < |A|$, we have $\mathrm{Inc}(A \times A) = |A|$. Finally, $\mathrm{Inc}(A \oplus B) = \max(|A|, |B|)$ if $|A|, |B| \geq 4$, since $A \oplus C \cong A \times A$ if $|C| = 4$.

Concerning derived functions of incomparability, we mention only a problem concerning (implicitly) $\mathrm{Inc}_{\mathrm{H}-}$:

Problem 53. *Is there an uncountable BA A with countable incomparability and $|B| > \omega$ for every infinite homomorphic image B of A?*

Now we turn to connections with our other functions. From the same Theorem 4.25 used above, it follows that $\mathrm{d}A \leq \mathrm{Inc}A$ for any infinite BA A; hence $\mathrm{hd}A \leq \mathrm{Inc}A$, by an easy argument. An example in which they are different is the interval algebra A on the reals. In fact, $\mathrm{hd}A = \omega$ by Theorem 14.1, and an incomparable set of size 2^ω is provided by

$$\{[0, r) \cup [1 + r, 2) : r \in (0, 1)\}.$$

Much effort has been put into constructing BA's A in which $\mathrm{Inc}A < |A|$. An example in ZFC of such an algebra is an algebra of Bonnet, Shelah [85]; the construction is too lengthy to be included here. Their algebra is an interval algebra, and has power $\mathrm{cf}(2^\omega)$. Rubin's algebra [83] is another example (constructed assuming $\diamondsuit$). Baumgartner [80] showed that it is consistent to have MA, $2^\omega = \omega_2$, and every uncountable BA has an uncountable incomparable subset.

We give a construction, using $\diamondsuit$, of a BA A with $\mathrm{Inc}A < |A|$; the example is due to Baumgartner, Komjath [81], and settles another question which is of interest. It depends on some lemmas. For brevity, let A be a denumerable atomless subalgebra of $\mathcal{P}\omega$, and let I be a maximal ideal in A. We consider a partial ordering $P = \{(a, b) : a \in I, b \in A \backslash I, a \subseteq b\}$, ordered by: $(a, b) \preceq (c, d)$ iff $a \supseteq c$ and $b \subseteq d$.

Lemma 15.2. *The following sets are dense in P:*

(i) For each $m \subseteq I$, the set $D_1 m \stackrel{\mathrm{def}}{=} \{(a, b) \in P :$ either $\forall c \in m(c \not\subseteq b)$ or $\exists c \in m(c \subseteq a)\}$.

(ii) For each $c \in A$, the set $D_2 c \stackrel{\mathrm{def}}{=} \{(a, b) \in P : \neg(a \subseteq c \subseteq b\})$.

(iii) For each $c \in I$, the set $D_3 c \stackrel{\mathrm{def}}{=} \{(a, b) \in P : c \subseteq a \cup (\omega \backslash b)\}$.

(iv) The set $D_4 \stackrel{\mathrm{def}}{=} \{(a, b) \in P : a \neq 0\}$.

Proof. Suppose $(a, b) \in P$. (i) If $\forall c \in m(c \not\subseteq b)$, then $(a, b) \in D_1 m$, as desired. Otherwise there is a $c \in m$ such that $c \subseteq b$, and then $(a \cup c, b) \in P$, $(a \cup c, b) \preceq (a, b)$, and $(a \cup c, b) \in D_1 m$, as desired.

(ii) If $a \subseteq c \subseteq b$, then there are two cases: Case 1. $c \in I$. Thus $c \subset b$. Choose disjoint non-empty d, e such that $b \backslash c = d \cup e$. Since $1 \notin I$, one of d, e is in I; say $d \in I$. Then $(a \cup d, b) \in P$, $(a \cup d, b) \preceq (a, b)$, and $(a \cup d, b) \in D_2 c$. Case 2. $c \notin I$. We can similarly find $d \subset (c \backslash a)$ such that $d \notin I$; then $(a, a \cup d)$ is the desired element showing that $D_2 c$ is dense.

(iii) The element $(a \cup (c \cap b), b)$ shows that $D_3 c$ is dense.

(iv) If $a \neq 0$, we are through. Otherwise choose non-empty disjoint c, d such that $c \cup d = b$. One of c, d, say c, is in I; then (c, b) is as desired. □

Lemma 15.3. *Suppose that D is dense in P. Then so are the following sets:*

(i) $SD \overset{\text{def}}{=} \{(a, b) \in P : (\omega \backslash b, \omega \backslash a) \in D\}$.

(ii) For $e, f \in I$, $T(D, e, f) \overset{\text{def}}{=} \{(a, b) \in P : ((a \backslash e) \cup f, (b \backslash e) \cup f) \in D\}$.

PROOF. Suppose $(a, b) \in P$. (i) We can choose $(c, d) \preceq (\omega \backslash b, \omega \backslash a)$ so that $(c, d) \in D$. Thus $\omega \backslash b \subseteq c$ and $d \subseteq \omega \backslash a$, so $\omega \backslash c \subseteq b$ and $a \subseteq \omega \backslash d$, which shows that $(\omega \backslash d, \omega \backslash c)$ is a desired element.

(ii) By Lemma 15.2(iii), choose $(a', b') \preceq (a, b)$ such that $e \cup f \subseteq a' \cup (\omega \backslash b')$. Let $e' = a' \cap e$ and $f' = a' \cap f$. By density, choose $(x, y) \in D$ such that $(x, y) \preceq ((a' \backslash e) \cup f, (b' \backslash e) \cup f)$. Now let

$$a'' = (x \backslash (e \cup f)) \cup e' \cup f' \text{ and } b'' = (y \backslash (e \cup f)) \cup e' \cup f'.$$

It is easy to check that $(a'', b'') \in T(D, e, f)$ and $(a'', b'') \preceq (a, b)$. □

Now suppose that M is a countable collection of subsets of I; then we let $\mathcal{D}M$ be the smallest collection of dense sets in P such that

(1) every set $D_1 m$, $D_2 c$, $D_3 e$, D_4 is in $\mathcal{D}M$ for $m \in M$, $c \in A$, $e \in I$;
(2) if $D \in \mathcal{D}M$ and $e, f \in I$, then $SD, T(D, e, f) \in \mathcal{D}M$.

A subset $x \subseteq \omega$ is *M−generic* if for all $D \in \mathcal{D}M$ there is an $(a, b) \in D$ such that $a \subseteq x \subseteq b$. Note that because $D_2 c \in \mathcal{D}M$ for all $c \in A$ we have $x \notin A$ in such a case.

Lemma 15.4. *For every M as above, there is a subset $x \subseteq \omega$ which is M-generic.*

PROOF. Let $\langle D_0, D_1 \ldots \rangle$ enumerate all members of $\mathcal{D}M$. Now we define (a_0, b_0), (a_1, b_1), ... by induction: $a_0 = 0$ and $b_0 = \omega$. Having defined (a_i, b_i), choose (a_{i+1}, b_{i+1}) so that $(a_{i+1}, b_{i+1}) \preceq (a_i, b_i)$ and $(a_{i+1}, b_{i+1}) \in D_i$. Let $x = \bigcup_{i<\omega} a_i$. Clearly x is as desired. □

Lemma 15.5. *Let x be M-generic and set $B = \langle A \cup \{x\} \rangle$. Then every element of $B \backslash A$ is M-generic. B is atomless, and $B \supset A$. Moreover, for any $a \in I$ we have $x \cap a \in I$ and $a \backslash x \in I$.*

PROOF. First we prove the final statement. In fact, choose $(c,d) \in D_3a$ so that $c \subseteq x \subseteq d$. Thus $a \subseteq c \cup (\omega\backslash d)$. It follows that $a \cap x \subseteq c \cap a \subseteq a \cap x$, as desired. And $a\backslash x \subseteq a\backslash d \subseteq a\backslash x$, as desired.

Now we claim:

(1) Every element of $B\backslash A$ has one of the two forms $(x\backslash e) \cup f$ or $((\omega\backslash x)\backslash e) \cup f$ for some $e, f \in I$.

In fact, take any element y of $B\backslash A$; we can write it in the form $y = (e \cap x) \cup (f\backslash x)$, where $e, f \in A$. By the above we cannot have $e, f \in I$. Now $-y = ((\omega\backslash e) \cap x) \cup ((\omega\backslash f)\backslash x)$, so by the above we also cannot have $-e, -f \in I$. So we have two cases: Case 1. $e \in I$ and $f \notin I$. Then $y = ((\omega\backslash x)\backslash(\omega\backslash f)) \cup (e \cap x)$, which is in one of the desired forms. Case 2. $e \notin I$ and $f \in I$. Then $y = (x\backslash(\omega\backslash e)) \cup (f\backslash x)$, which again is in one of the desired forms. So (1) holds.

Next

(2) If $e, f \in I$, then $(x\backslash e) \cup f$ is M-generic.

For, given $D \in \mathcal{D}M$, we also have $T(D, e, f) \in \mathcal{D}M$, and hence there is an $(a, b) \in T(D, e, f)$ such that $a \subseteq x \subseteq b$. Then $(a\backslash e) \cup f \subseteq (x\backslash e) \cup f \subseteq (b\backslash e) \cup f$, and $((a\backslash e) \cup f, (b\backslash e) \cup f) \in D$, as desired.

Finally, since $\mathcal{D}M$ is closed under the operation S, it follows easily that $\omega\backslash x$ is M-generic. From (1) and (2) the first conclusion of the lemma now follows.

That B is atomless follows from what has already been shown, plus the fact that $D_4 \in \mathcal{D}M$. And B is a proper extension of A since $D_2c \in \mathcal{D}M$ for every $c \in A$. □

Lemma 15.6. *Under the hypotheses of Lemma 15.5,*
(i) for all $a \in I$ and all $b \in B$, if $b \subseteq a$ then $b \in A$.
(ii) $I' \stackrel{\text{def}}{=} \langle I \cup \{x\}\rangle^{\text{Id}}$ is a maximal ideal in B.

PROOF. (i): By Lemma 15.5, if $b \notin A$ then b is M-generic, so by the last comment of Lemma 15.5, $a \cap b \in A$; so, of course, $b \not\subseteq a$.

(ii): First suppose that I' is not proper; write $\omega = x \cup a$ with $a \in I$. Thus $\omega\backslash a \subseteq x$ so, since $a\backslash x \in A$ by Lemma 15.5, its complement is also in A, and $x = (\omega\backslash a) \cup x \in A$, contradiction. So, I' is a proper ideal.

Since $u \in I'$ or $-u \in I'$ for every $u \in A \cup \{x\}$, and $A \cup \{x\}$ generates B, it follows that I' is maximal. □

EXAMPLE 15.7. (The Baumgartner-Komjath algebra.) We construct a BA A such that $\text{Inc}A = \omega = \text{Length}A$, while $\chi A = \omega_1$. $\diamondsuit$ is assumed.

Let $\langle S_\alpha : \alpha \in \omega_1\rangle$ be a $\diamondsuit$-sequence, and let $\langle a_\alpha : \alpha \in \omega_1\rangle$ be a one-one enumeration of $\mathcal{P}\omega$. For each $\beta < \omega_1$ let $m_\beta = \{a_\alpha : \alpha \in S_\beta\}$.

We define sequences $\langle A_\alpha : \alpha < \omega_1\rangle$, $\langle I_\alpha : \alpha < \omega_1\rangle$, $\langle M_\alpha : \alpha < \omega_1\rangle$ by induction, as follows. Let A_0 be a denumerable atomless subalgebra of $\mathcal{P}\omega$, and let I_0 be a maximal ideal in A_0. If we have defined a denumerable atomless subalgebra A_α of $\mathcal{P}\omega$ and a maximal ideal I_α of A_α, we let $M_\alpha = \{m_\beta : \beta \leq \alpha$, and $m_\beta \subseteq I_\alpha\}$. Let x_α be M_α-generic (with respect to A_α and I_α), and let $A_{\alpha+1} = \langle A_\alpha \cup \{x_\alpha\}\rangle$ and $I_{\alpha+1} = \langle I_\alpha \cup \{x_\alpha\}\rangle^{\mathrm{Id}}$. For α a limit ordinal let $A_\alpha = \bigcup_{\beta<\alpha} A_\beta$ and $I_\alpha = \bigcup_{\beta<\alpha} I_\beta$. Finally, let $A = \bigcup_{\alpha<\omega_1} A_\alpha$ and $I = \bigcup_{\alpha<\omega_1} I_\alpha$.

From the above lemmas it is clear that each A_α is atomless, and hence A is atomless. Furthermore, I is a maximal ideal of A, and $|A \upharpoonright a| = \omega$ for every $a \in I$. Moreover, $|A| = \omega_1$. Clearly the filter dual to I has character ω_1, so $\chi A = \omega_1$.

Suppose that m is an uncountable incomparable set. Now trivially $a \subseteq b$ iff $\omega\backslash b \subseteq \omega\backslash a$, so we may assume that $m \subseteq I$. Let $S = \{\alpha : a_\alpha \in m\}$, and let Z be the set of all α satisfying the following two conditions:

(1) $\{a_\beta : \beta \in S \cap \alpha\} = m \cap A_\alpha$.

(2) For all $b \in A_\alpha\backslash I_\alpha$, if there is a $c \in m$ such that $c \subseteq b$, then there is a $\beta \in S \cap \alpha$ such that $a_\beta \subseteq b$.

Clearly Z is club. Hence by the $\diamondsuit$ property, choose $\alpha \in Z$ such that $S_\alpha = S \cap \alpha$.

(3) $m_\alpha \subseteq A_\alpha \cap m$.

For, let $x \in m_\alpha$. Say $x = a_\beta$ with $\beta \in S_\alpha = S \cap \alpha$. Since $\alpha \in Z$, we get $x \in m \cap A_\alpha$.

For each $c \in A\backslash A_0$ let ρc be the least β such that $c \in A_{\beta+1}\backslash A_\beta$. Now pick $c \in m\backslash m_\alpha$. Write $\rho c = \beta$. Now $\beta \geq \alpha$: if $\beta < \alpha$, then $c \in A_\alpha \cap m$, so, since $\alpha \in Z$, we have $c = a_\gamma$ for some $\gamma \in S \cap \alpha = S_\alpha$, so $c \in m_\alpha$, contradiction.

Since c is M_β-generic (with respect to A_β and I_β) and $m_\alpha \in M_\beta$, there is a $(a, b) \in D_1 m_\alpha$ such that $a \subseteq c \subseteq b$. By the definition of $D_1 m_\alpha$ and since c is not comparable with any element of m_α, we must have $\forall c' \in m_\alpha(c' \not\subseteq b)$. Choose b' with $\rho b'$ minimum such that $c \subseteq b'$ and $\forall c' \in m_\alpha(c' \not\subseteq b')$.

Now $\rho b' \geq \alpha$: suppose not. Then $b' \in A_\alpha\backslash I_\alpha$. Now for any γ, if $\gamma \in S \cap \alpha$ then $\gamma \in S_\alpha$, $a_\gamma \in m_\alpha$, and $a_\gamma \not\subseteq b'$. This contradicts (2) for α.

Say $\rho b' = \gamma \geq \alpha$. Now b' is M_γ-generic (with respect to A_γ and I_γ) and $m_\alpha \in M_\gamma$, so there is a $(a'', b'') \in D_1 m_\alpha$ such that $a'' \subseteq b' \subseteq b''$; note that $a'', b'' \in A_\gamma$. For any $c' \in m_\alpha$ we have $c' \not\subseteq a''$; hence by the definition of $D_1 m_\alpha$ we have $\forall c' \in m_\alpha(c' \not\subseteq b'')$. Since $c \subseteq b''$ and $\rho b'' < \gamma$, this contradicts the minimality of $\rho b'$. Thus we have shown that A has no uncountable incomparable set.

If C is an uncountable chain in A, we may assume that $C \subseteq I$. We define $\langle c_\alpha : \alpha < \omega_1\rangle$. Suppose $c_\beta \in C$ has been constructed for all $\beta < \alpha$. Say $\{c_\beta : \beta < \alpha\} \subseteq A_\gamma$. Then $\{c : c \in C$ and $c \leq c_\beta$ for some $\beta < \alpha\} \subseteq A_\gamma$ by Lemma 15.6(i). So, we can choose $c_\alpha \in C$ such that $c_\beta < c_\alpha$ for all $\beta < \alpha$.

The sequence so constructed shows that Depth$A = \omega_1$; hence Inc$A = \omega_1$, contradiction. □

Problem 54. *Can one construct in ZFC a BA A such that* Inc$A < \chi A$*?*

We should mention in connection with Example 15.7 that Shelah [80], and independently van Wesep, showed that it is consistent to have 2^ω arbitrarily large and to have a BA of size 2^ω whose length and incomparability are countable.

We conclude this section with some remarks about incomparability in subalgebras of interval algebras. By Theorem 15.22 of Part I of the BA handbook, if κ is uncountable and regular, and B is a subalgebra of an interval algebra and $|B| = \kappa$, then B has a chain or incomparable subset of size κ. M. Bekkali has shown that it is consistent that this no longer holds for singular cardinals.

16. HEREDITARY COFINALITY

Theorem 16.1. h-cof$A = \sup\{|T| : T \subseteq A,\ T$ *well-founded*$\}$.

PROOF. Call these two cardinals κ_0 and κ_1. Suppose that $\kappa_1 < \kappa_0$. Let X be a subset of A having no cofinal subset of power $\leq \kappa_1$. We construct elements $\langle x_\alpha : \alpha < \kappa_1^+\rangle$ by induction: if x_α has been defined for all $\alpha < \beta$, with $\beta < \kappa_1^+$, then $\{x_\alpha : \alpha < \beta\}$ is not cofinal in X, so there is an $x_\beta \in X$ such that $x_\beta \not\leq x_\alpha$ for all $\alpha < \beta$. This finishes the construction. Now $\{x_\alpha : \alpha < \kappa_1^+\}$ is not well-founded, so there exist $\alpha 0, \alpha 1, \ldots < \kappa_1^+$ such that $x_{\alpha 0} > x_{\alpha 1} > \ldots$. Choose $i < j$ such that $\alpha i < \alpha j$. Then $x_{\alpha j} < x_{\alpha i}$ is a contradiction.

Suppose $\kappa_0 < \kappa_1$. Let T be a well-founded subset of A of power κ_0^+. If T has κ_0^+ incomparable elements, this is a contradiction. So T has $\geq \kappa_0^+$ levels. Let T' consist of all elements of T of level $< \kappa_0^+$. Let $X \subseteq T'$ be a cofinal subset of T' of cardinality $\leq \kappa_0$. Then choose $a \in T'$ of level greater than the levels of all members of X; clearly this is impossible. □

We have not investigated the behaviour of h-cof under algebraic operations.

Concerning relationships to our other functions, the main facts are that Inc$A \leq$ h-cof$A \leq |A|$ and hL$A \leq$ h-cof$A$. To see that hL$A \leq$ h-cof$A$, suppose that $\langle F_\alpha : \alpha < \kappa\rangle$ is right-separated. For every $\alpha < \kappa$ choose $x_\alpha \in F_\alpha$ such that for all $\beta < \kappa$, if $x_\alpha \in F_\beta$, then $\beta \leq \alpha$. Then $\{x_\alpha : \alpha < \kappa\}$ is well-founded: suppose $x_{\alpha 0} > x_{\alpha 1} > \ldots$. Choose $i < j$ with $\alpha i < \alpha j$. Then $x_{\alpha i} \in F_{\alpha j}$, contradiction. It is obvious that Inc$A \leq$ h-cof$A \leq |A|$. An example in which hL$A <$ h-cof$A$ is provided by the interval algebra A on the reals. In fact, in Lemma 1.6 we showed that hL$A = \omega$, and in Section 15 we showed that Inc$A = 2^\omega$, and hence h-cof$A = 2^\omega$. Since $\chi A \leq$ hL$A \leq$ h-cof$A$, the Baumgartner-Komjath algebra of section 15 provides an example where Inc$A <$ h-cof$A$.

Problem 55. *Can one construct in ZFC a BA A with the property that* Inc$A <$ h-cof$A$*?*

To complete this picture, it remains to provide an example in which h-cof is less than cardinality. An example of this, in ZFC, is the algebra of Bonnet and Shelah [85]. Since, as mentioned before, the construction of that algebra requires too much space for us to give it here, we are going to describe a different construction, the algebra of Rubin [83]. It requires $\diamondsuit$, but it will be used later too. It is relevant to many of our functions and problems.

EXAMPLE 16.2. Rubin's construction is not direct, but goes by way of more general considerations. Let A be a BA. A *configuration* for A is, for some $n \in \omega$, an $(n+3)$-tuple $\langle a, c_1, c_2, b_1, \ldots, b_n\rangle$ such that $a, b_1, \ldots, b_n$ are pairwise disjoint, each $b_i \neq 0$, $c_1 \subseteq a + \sum_{i=1}^n b_i$, $a + c_1 \leq c_2$, and $(c_2 - c_1) \cdot b_i \neq 0$ for all

$i = 1, \ldots, n$. (See Figure 16.3.) Now we call a subset P of A *nowhere dense for configurations in* A, for brevity nwdc in A, if for every $n \in \omega$ and all disjoint $a, b_1, \ldots, b_n$ with each $b_i \neq 0$, there exist c_1, c_2 such that $\langle a, c_1, c_2, b_1, \ldots, b_n\rangle$ is a configuration and $P \cap (c_1, c_2) = 0$. Rubin's theorem that we are aiming for says that, assuming $\diamondsuit$, there is an atomless BA A of power ω_1 such that every set which is nwdc in A is countable. Before proceeding to the proof of this theorem, let us check that for such an algebra we have h-cof$A = \omega$. Suppose that P is an uncountable subset of A. Thus P is not nwdc, so we get $n \in \omega$ and disjoint $a, b_1, \ldots, b_n$ with each $b_i \neq 0$ such that

(1) For all c_1, c_2, if $\langle a, c_1, c_2, b_1, \ldots, b_n\rangle$ is a configuration it follows that $P \cap (c_1, c_2) \neq 0$.

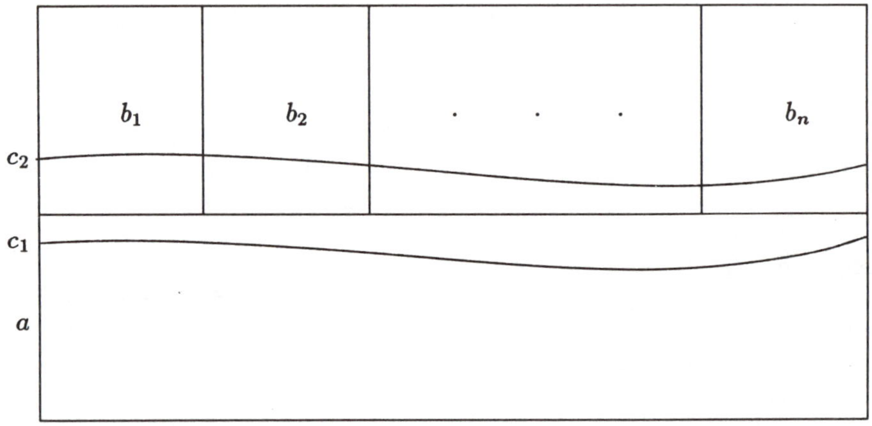

Figure 16.3

Now if we start with a and an element c_0 of A such that $a \leq c_0$ and $c_0 \cdot b_i \neq 0$ for all $i = 1, \ldots, n$, we can use (1) in a clear fashion to produce elements $c_1, c_2, \ldots$ of P such that $c_1 > c_2 > \cdots$, which means that P is not well-founded. This shows that h-cof$A = \omega$.

To do the actual construction leading to Rubin's theorem, we need another definition and two lemmas. Let A be a BA and assume that $P \subseteq B \subseteq A$. We say that *$P$ is B-nowhere dense for configurations in A*, for brevity P is B-nwdc in A, if for every $n \in \omega$ and all disjoint $a, b_1, \ldots, b_n \in A$ with each $b_i \neq 0$, there exist $c_1, c_2 \in B$ such that $\langle a, c_1, c_2, b_1, \ldots, b_n\rangle$ is a configuration and $P \cap (c_1, c_2) = 0$. Thus to say that P is nwdc in A is the same as saying that P is A-nwdc in A.

An important tool in the construction is the general notion of the free extension $A(x)$ of a BA A obtained by adjoining an element x (and other elements necessary when it is adjoined); this is the free product of A with a BA with four elements $0, x, -x, 1$. We need only this fact about this procedure:

Lemma 16.4. *Let A be a BA and $A(x)$ the free extension of A by an element x. Suppose that $\langle a_i : i \in I\rangle$ is a system of disjoint elements of A, $\langle b_i : i \in I\rangle$ is another system of elements of A, and $b_i \leq a_i$ for all $i \in I$. Let $I = \langle\{(a_i \cdot x)\triangle b_i : i \in I\rangle^{\mathrm{Id}}$, and let k be the natural homomorphism from $A(x)$ onto $A(x)/I$. Then $k \restriction A$ is one-one.*

PROOF. Suppose $kc = 0$, with $c \in A$. Then there exist $i(0), \ldots, i(m) \in I$ such that

$$c \leq (a_{i(0)} \cdot x)\triangle b_{i(0)} + \cdots + (a_{i(m)} \cdot x)\triangle b_{i(m)};$$

letting f be a homomorphism of $A(x)$ into A such that f is the identity on A and $fx = b_{i(0)} + \cdots + b_{i(m)}$, we infer that $c = 0$, as desired. □

Note that the effect of the ideal I in Lemma 16.4 is to subject x to the condition that $x \cdot a_i = b_i$ for all $i \in \omega$. The following is the main lemma leading to Rubin's construction.

Lemma 16.5. *Let A be a denumerable atomless BA. Assume that for each $i < \omega$ we have $P_i \subseteq B_i \subseteq A$ and P_i is B_i-nwdc for A. Then there is a proper extension A' of A such that A is dense in A' and P_i is B_i-nwdc for A' for all $i < \omega$.*

PROOF. Let $A(x)$ be a free extension of A by an element x; we shall obtain the desired algebra A' by the procedure of Lemma 16.4; thus we will let $A' = A(x)/I$, with I specified implicitly by defining a_j's and b_j's. Let $\langle s_n : n < \omega\rangle$ be an enumeration of the following set:

$$\{\langle 0, a\rangle : a \in A\} \cup \{\langle 1, a, b, c\rangle : a, b, c \text{ are disjoint elements of } A\} \cup$$
$$\{\langle 2, a, b, c, b_1, \ldots, b_k, i\rangle : a, b, c \text{ are disjoint elements of } A,$$
$$b_1, \ldots, b_n \text{ are disjoint non-zero elements of } A, \text{ and } i < \omega\}.$$

As we shall see, $\langle s_n : n < \omega\rangle$ is a list of things to be done in coming up with the ideal I. We will take care of the objects s_i by induction on i. Suppose that we have already taken care of s_i for $i < n$, having constructed a_j and b_j for this purpose, $j \in J$, so that J is a finite set, $b_j \leq a_j$ for all $j \in J$, the a_j's are pairwise disjoint, and $\sum_{j\in J} a_j < 1$. Let $u = \sum_{j\in J} a_j$ and $v = \sum_{j\in J} b_j$. We want to take care of s_n so that these conditions (called the "list conditions") will still be satisfied. Note that under I, $x \cdot u$ will be equivalent to v, and $-x \cdot u$ will be equivalent to $u \cdot -v$. Now we consider three cases, depending upon whether the first term of s_n is 0,1, or 2.

Case 1. The first term of s_n is 0; say $s_n = \langle 0, a\rangle$, where $a \in A$. We want to add new elements a_k and b_k to our lists in order to insure that $[x] \neq [a]$

in $A(x)/I$, where in general $[z]$ denotes the equivalence class of $z \in A(x)$ with respect to I. Thus the fact that this case is taken care of for all s_n of this type in our list will insure merely that $A(x)/I$ is a proper extension of A. If $a + u \neq 1$, choose e so that $0 < e < -(a+u)$, and set $a_k = b_k = e$. Then in the end we will have $(e \cdot x)\Delta e \in I$, hence $0 < [e] \leq [x]$, and $[e] \cdot [a] = 0$, so $[x] \neq [a]$. Clearly the list conditions still hold. Now suppose that $a + u = 1$. Thus $-u \leq a$, and $-u \neq 0$. Choose e with $0 < e < -u$. Let $a_k = e$ and $b_k = 0$. Then in the end we will have $e \cdot x \in I$, hence $[e] \cdot [x] = 0$, and $0 < [e] \leq [a]$, so $[a] \neq [x]$. And again the list conditions hold.

Case 2. The first term of s_n is 1; say $s_n = \langle 1, a, b, c \rangle$, where a, b, c are disjoint elements of A. We consider the element $t \stackrel{\text{def}}{=} a + b \cdot x + c \cdot -x$; we want to fix things so that if $[t]$ is non-zero then there will be some element $u \in A$ such that $0 < [u] \leq [t]$. This will insure that A will be dense in $A(x)/I$. Now

$$t = a + b \cdot x \cdot u + b \cdot x \cdot -u + c \cdot -x \cdot u + c \cdot -x \cdot -u,$$

and under I this is equivalent to

$$a + b \cdot v + u \cdot -v \cdot c + b \cdot -u \cdot x + c \cdot -u \cdot -x.$$

Let $a' = a + b \cdot v + u \cdot -v \cdot c$, $b' = b \cdot -u$, $c' = c \cdot -u$; thus a', b', c' are disjoint. If $a' \neq 0$, we don't need to add anything to our lists. Suppose that $b' \neq 0$. Then choose e with $0 < e < b'$, and add a_k, b_k to our lists, where $a_k = b_k = e$; this assures that $[e] \leq [x]$, hence $0 < [e] \leq [t]$; clearly the list conditions hold. If $c' \neq 0$ a similar procedure works. Finally, if $a' = b' = c' = 0$, then $[t] = 0$, and again we do not need to add anything.

Case 3. The first term of s_n is 2; say $s_n = \langle 2, a, b, c, b_1, \ldots, b_k, i \rangle$, where a, b, c are disjoint elements of A, $b_1, \ldots, b_k$ are disjoint non-zero elements of A, and $i < \omega$. Let t be as in case 1. Case 3 is the crucial case, and here we will do one of three things: (1) make t equivalent to an element of A; (2) make sure that $[t] \cdot [b_j] \neq 0$ for some $j = 1, \ldots, k$; (3) find $c_1, c_2 \in B_i$ with $P_i \cap (c_1, c_2) = 0$ so that $\langle [t], [c_1], [c_2], [b_1], \ldots, [b_k] \rangle$ is a configuration. Thus this step will assure in the end that P_i is B_i-nwdc for A', since (1) enables us to apply the assumption of the lemma, (2) means that the desired condition holds vacuously, and (3) yields what is desired. Note that denseness of A in A' enables us to take all b_i in A rather than in A'. Let a', b', c' be as above. If $\sum_{j=1}^{k} b_i \cdot a' \neq 0$, then (2) will automatically hold, and we do not need to add anything to our lists. If there is a j, $1 \leq j \leq k$, such that $b_j \cdot b' \neq 0$, let e be such that $0 < e < b_j \cdot b'$, and adjoin a_l, b_l to our lists, where $a_l = b_l = e$; then we will have $[e] \leq [x]$, and $[b_j] \cdot [t] \neq 0$, which means that (2) holds—and the list conditions are ok. Similarly if $b_j \cdot c' \neq 0$ for some j. If $b' + c' = 0$, then $[t] = [a']$, i.e., (1) holds. Thus we are left with the essential situation: $b' + c' \neq 0$, and $(a' + b' + c') \cdot b_j = 0$ for all $j = 1, \ldots, k$. First of all we use the fact that P_i is B_i-nwdc for A, applied to $a', b'+c', b_1, \ldots, b_k$, to get $c_1, c_2 \in B_i$ such that $P_i \cap (c_1, c_2) = 0$ and $\langle a', c_1, c_2, b'+c', b_1, \ldots, b_k\rangle$ is a configuration. This time we add elements a_l, a_m, b_l, b_m to our lists, where $a_l = c_1 \cdot (b'+c')$, $b_l = c_1 \cdot b'$, $a_m = (b'+c') \cdot -c_2$, and $b_m = c' \cdot -c_2$. Clearly $a_l \cdot a_m = 0$ and both elements are disjoint from previous a_j's. Obviously $b_l \leq a_l$ and $b_m \leq a_m$. Next, since $\langle a', c_1, c_2, b'+c', b_1, \ldots, b+k\rangle$ is a configuration we have $c_2 \cdot -c_1 \cdot (b'+c') \neq 0$, and since this element is disjoint from all previous a_j's as well as from a_l and a_m it follows that $u + a_l + a_m < 1$. Thus the list conditions hold. It remains only to show that in the end $\langle [t], [c_1], [c_2], [b_1], \ldots, [b_k]\rangle$ is a configuration. The only things not obvious are that $[c_1] \leq [t + \sum_{j=1}^k b_j]$ and $[t] \leq [c_2]$. Since $\langle a', c_1, c_2, b'+c', b_1, \ldots, b_k\rangle$ is a configuration, we have $c_1 \leq a' + b' + c' + b_1 + \cdots + b_k$. Hence in order to show that $[c_1] \leq [t + \sum_{j=1}^k b_j]$, it suffices to prove that $[c_1 \cdot (b'+c')] \leq [t]$, and that is done as follows. First note that our added elements a_l, a_m, b_l, b_m assure that $[x \cdot c_1 \cdot (b'+c')] = [c_1 \cdot b']$ and $[x \cdot (b'+c') \cdot -c_2] = [c' \cdot -c_2]$, hence $[c_1 \cdot b'] \leq [x]$ and $[c' \cdot -c_2] \leq [x]$.

$$\begin{aligned}
[t] &\geq [b' \cdot x + c' \cdot -x] \\
&\geq [b' \cdot c_1 \cdot x + c' \cdot -x \cdot c_1] \\
&= [c_1 \cdot b' + c' \cdot c_1 \cdot -x] \\
&= [c_1 \cdot b' + (c' \cdot c_1) \cdot -(c' \cdot c_1 \cdot x)] \\
&\geq [c_1 \cdot b' + (c' \cdot c_1) \cdot -((b'+c') \cdot c_1 \cdot x)] \\
&= [c_1 \cdot b' + (c' \cdot c_1) \cdot -(c_1 \cdot b')] \\
&= [c_1 \cdot b' + c_1 \cdot c'] \\
&= [c_1 \cdot (b'+c')].
\end{aligned}$$

To show that $[t] \leq [c_2]$, it suffices to show that $[t \cdot (b'+c')] \leq [c_2]$, and that is done like this:

$$\begin{aligned}
[t \cdot (b'+c')] &= [t \cdot (b'+c') \cdot c_2 + t \cdot (b'+c') \cdot -c_2] \\
&\leq [c_2 + t \cdot (b'+c') \cdot -c_2] \\
&= [c_2 + b' \cdot x \cdot (b'+c') \cdot -c_2 + c' \cdot -x \cdot (b'+c') \cdot -c_2] \\
&= [c_2 + b' \cdot c' \cdot -c_2 + c' \cdot -c_2 \cdot -x] \\
&= [c_2].
\end{aligned}$$

This completes the construction and the proof. □

EXAMPLE 16.2 (Conclusion). Recall that we are trying to construct, using $\diamondsuit$, an atomless BA A of power ω_1 such that every nwdc subset of A is countable. We shall define by induction an increasing sequence $\langle A_\alpha : \alpha < \omega_1,\ \alpha \text{ limit}\rangle$ of countable BA's, and a sequence $\langle P_\alpha : \alpha < \omega_1,\ \alpha \text{ limit}\rangle$, such that: the universe of A_α is α; A_ω is atomless and is dense in A_α for all limit $\alpha < \omega_1$; $P_\alpha \subseteq A_\alpha$ for all limit $\alpha < \omega_1$, and P_α is A_α-nwdc for A_β whenever α, β are limit ordinals $< \omega_1$ with $\alpha \leq \beta$.

Let $\langle S_\alpha : \alpha < \omega_1 \rangle$ be a $\diamondsuit$-sequence. Let A_ω be a denumerable atomless BA. If λ is a limit of limit ordinals, $\lambda < \omega_1$, let $A_\lambda = \bigcup_{\alpha<\lambda} A_\alpha$. If S_λ is nwdc for A_λ, let $P_\lambda = S_\lambda$, and let $P_\lambda = 0$ otherwise. Now suppose that α is a limit ordinal $< \omega_1$, and A_β and P_β have been defined for all limit ordinals $\beta \leq \alpha$. By Lemma 16.5 let $A_{\alpha+\omega}$ be a BA with universe $\alpha + \omega$ such that A_α is dense in $A_{\alpha+\omega}$ and P_β is A_β-nwdc for $A_{\alpha+\omega}$ for all limit $\beta \leq \alpha$. And again choose $P_{\alpha+\omega} = S_{\alpha+\omega}$ if $S_{\alpha+\omega}$ is nwdc for $A_{\alpha+\omega}$, and let it be 0 otherwise. This completes the inductive definition. Let $A = \bigcup\{A_\alpha : \alpha \text{ limit}, \alpha < \omega_1\}$.

Clearly A is atomless and of power ω_1. Now suppose, in order to get a contradiction, that P is an uncountable nwdc subset of A. Let

$$F = \{\alpha : \alpha < \omega_1, \alpha \text{ limit, and } (A_\alpha, P \cap \alpha) \preceq_{\text{ee}} (A, P).$$

Here $\preceq_{\text{ee}}$ means "elementary substructure". Clearly F is club in ω_1. Now by the $\diamondsuit$-property, the set $S \stackrel{\text{def}}{=} \{\alpha < \omega_1 : \alpha \cap P = S_\alpha\}$ is stationary, so we can choose $\alpha \in F \cap S$. Clearly nwdc can be expressed by a set of first-order formulas; so $P \cap \alpha$ is nwdc in A_α. Since $P \cap \alpha = S_\alpha$, the construction then says that $P_\alpha = S_\alpha$. Since P is uncountable, choose $a \in P \backslash P_\alpha$, and then choose $c_1, c_2 \in A_\alpha$ so that $\langle a, c_1, c_2 \rangle$ is a configuration (this means just so that $c_1 \leq a \leq c_2$) and $P_\alpha \cap (c_1, c_2) = 0$; this is possible, since if $a \in A_\beta$ with $\alpha \leq \beta$, then P_α is A_α-nwdc for A_β by the construction. But then we have

$$(A_\alpha, P_\alpha) \models \forall x[P(x) \rightarrow x \notin (c_1, c_2)];$$
$$(A, P) \models P(a) \wedge x \in (c_1, c_2).$$

This contradicts the fact that $(A_\alpha, P_\alpha) \preceq_{\text{ee}} (A, P)$. □

17. NUMBER OF ULTRAFILTERS

This cardinal function is rather easy to describe, at least if we do not try to go into the detail that we did for cellularity, for example. If A is a subalgebra or homomorphic image of B, then $|\mathrm{Ult}A| \leq |\mathrm{Ult}B|$. For weak products we have $|\prod^{\mathrm{w}}_{i\in I} A_i| = \max(\omega, \sup_{i\in I}|\mathrm{Ult}A_i|)$. The situation for full products is more complicated:

$$\left|\mathrm{Ult}\left(\prod_{i\in I} A_i\right)\right| \leq 2^{2^{\kappa}},$$

where $\kappa = \sum_{i\in I} \mathrm{d}A_i$. This follows from the following two facts:

$$\prod_{i\in I} A_i \rightarrowtail \prod_{i\in I} \mathcal{P}(\mathrm{d}A_i) \cong \mathcal{P}\left(\dot{\bigcup}_{i\in I} \mathrm{d}A_i\right),$$

where "$\rightarrowtail$" means "is isomorphically embeddable in", and "$\dot{\bigcup}$" means "disjoint union". Next, clearly $|\mathrm{Ult} \oplus_{i\in I} A_i| = \prod_{i\in I} |\mathrm{Ult}A_i|$.

Concerning relationships to our other functions, we mention only that $|A| \leq |\mathrm{Ult}A|$ and $2^{\mathrm{Ind}A} \leq |\mathrm{Ult}A|$.

About $|\mathrm{Ult}A|$ for A in special classes of BA's: first recall from Theorem 17.10 of Part I of the BA handbook that $|\mathrm{Ult}A| = |A|$ for A superatomic. If A is not superatomic, then $|\mathrm{Ult}A| \geq 2^{\omega}$, since A has a denumerable atomless subalgebra B, and obviously $|\mathrm{Ult}B| = 2^{\omega}$.

18. NUMBER OF AUTOMORPHISMS

This cardinal function is not related very much to the preceding ones. To start with, we state some general facts about the size of automorphism groups in BA's; for proofs or references, see the chapter on automorphism groups in the BA handbook.

1. If A is denumerable, then $|\text{Aut}A| = 2^\omega$.
2. If $0 \neq m \in \omega$ and $\kappa > \omega$, then there is a BA A with $|A| = \kappa$ such that $|\text{Aut}A| = m!$.
3. If $|\text{Aut}A| < \omega$, then $|\text{Aut}A| = m!$ for some $m \in \omega$.
4. If MA and $|\text{Aut}A| = \omega$, then $|A| \geq 2^\omega$.
5. If $2^\omega \leq \kappa$, then there is a BA A such that $|\text{Aut}A| = \omega$ and $|A| = \kappa$.
6. If $\omega < \kappa \leq \lambda$, then there is a BA A with $|A| = \lambda$ and $|\text{Aut}A| = \kappa$.
7. If $\omega \leq \kappa$, then there is a BA A with $|A| = \kappa$ and $|\text{Aut}A| = 2^\kappa$.
8. Any BA can be embedded in a rigid BA.
9. Any BA can be embedded in a homogeneous BA.

Now we discuss algebraic operations on BA's vis-à-vis automorphism groups. If A is a subalgebra or homomorphic image of B, then $|\text{Aut}A|$ can vary in either direction from $|\text{Aut}B|$: embedding a rigid BA A into a homogeneous BA B, we get $|\text{Aut}A| < |\text{Aut}B|$, while if we embed a free BA A in a rigid BA B we get $|\text{Aut}A| > |\text{Aut}B|$; any rigid BA A is the homomorphic image of a free BA B, and then $|\text{Aut}A| < |\text{Aut}B|$; and finally, embed $A \overset{\text{def}}{=} \mathcal{P}\omega$ into a rigid BA B, and then extend the identity on A to a homomorphism from B onto A—this gives $|\text{Aut}A| > |\text{Aut}B|$.

Now we consider products. There are two fundamental, elementary facts here. First, $|A| \leq |\text{Aut}(A \times A)|$ for any BA A. This is easily seen by the following chains of isomorphisms, starting from any element $a \in A$ to produce an automorphism f_a of $A \times A$:

$$\begin{aligned} A \times A &\overset{g}{\cong} (A \restriction a) \times (A \restriction -a) \times (A \restriction a) \times (A \restriction -a) \\ &\overset{h}{\cong} (A \restriction a) \times (A \restriction -a) \times (A \restriction a) \times (A \restriction -a) \\ &\overset{g^{-1}}{\cong} A \times A, \end{aligned}$$

where g is the natural mapping and h interchanges the first and third factors, leaving the second and fourth fixed. If $a \neq b$, then $f_a \neq f_b$; in fact, say $a \not\leq b$; then $f_a(a, 0) = (0, a)$ while $f_b(a, 0) = (a \cdot -b, a \cdot b) \neq (0, a)$. This proves that $|A| \leq |\text{Aut}(A \times A)|$. The second fact is that the group $\text{Aut}A \times \text{Aut}B$ embeds isomorphically into $\text{Aut}(A \times B)$; an isomorphism F is defined like this, for any $f \in \text{Aut}A$, $g \in \text{Aut}B$, $a \in A$, $b \in B$: $(F(f, g))(a, b) = (fa, gb)$. Putting these two elementary facts together, we have $|A|$, $|\mathrm{Aut}A|$ both $\leq |\mathrm{Aut}(A \times A)|$. We do not know whether $<$ can be attained:

Problem 56. *Is there an infinite BA A such that* $|A|$ *and* $|\mathrm{Aut}A|$ *are both smaller than* $|\mathrm{Aut}(A \times A)|$*?*

For weak products, we have $\sup_{i\in I}|\mathrm{Aut}A_i| \leq |\mathrm{Aut}\left(\prod^{\mathrm{w}}_{i\in I} A_i\right)|$ by the above remarks; a similar statement holds for full products—in fact, the full direct product of groups $\prod_{i\in I}\mathrm{Aut}A_i$ is isomorphically embeddable in $\mathrm{Aut}\left(\prod_{i\in I} A_i\right)$.

The situation for free products is much like that for products. By Proposition 11.11 of the BA handbook, Part I, every automorphism of A extends to one of $A \oplus B$; so $|\mathrm{Aut}(A \oplus B)| \geq \max(|\mathrm{Aut}A|, |\mathrm{Aut}B|)$. And $|A| \leq |\mathrm{Aut}(A \oplus A)|$. In fact, choose $a \in A$ with $0 < a < 1$. Then $|(A \oplus A) \restriction (a \times -a)| = |A|$, $(A \oplus A) \restriction (a \times -a) \cong (A \oplus A) \restriction (-a \times a)$, and

$$A \oplus A \cong [(A \oplus A) \restriction (a \times -a)] \times [(A \oplus A) \restriction (-a \times a)] \times [(A \oplus A) \restriction c]$$

for some c, so our statement follows from the above considerations on products. But we have a problem similar to problem 56:

Problem 57. *Is there an infinite BA A such that* $|A|$ *and* $|\mathrm{Aut}A|$ *are both smaller than* $|\mathrm{Aut}(A \oplus A)|$*?*

As mentioned at the beginning of this section, $|\mathrm{Aut}A|$ is not strongly related to our previous cardinal functions. An example with the property that $|\mathrm{Aut}A| < \mathrm{Depth}A$ is provided by embedding the interval algebra on κ into a rigid BA A. A similar procedure can be applied for independence and π-character, and these three examples show similar things for all of our preceding functions. In the case of π-character it is convenient to use Theorem 9.6 to verify that a ccc rigid BA of high cardinality has large π-character. And recall from the section on incomparability that if A is cardinality-homogeneous and has no incomparable subset of size $|A|$, then A is rigid.

Concerning automorphisms of special kinds of BA's, we mention only the fact that $|\mathrm{Aut}A| = 2^\kappa$ for A the interval algebra on κ. In fact, every automorphism of A is induced by a permutation of κ; so we just need to describe 2^κ permutations of κ that give rise to automorphisms of A. For each $\alpha < \kappa$ we can consider the transposition $(\omega\cdot\alpha+1, \omega\cdot\alpha+2)$. For each $\varepsilon \in {}^\kappa 2$ let f_ε be the permutation of κ which, on the interval $[\omega\cdot\alpha, \omega\cdot\alpha+\omega)$, is this transposition if $\varepsilon\alpha = 1$, and is the identity there otherwise. It is easy to see that the function on A induced by f_ε maps into A, and hence is an automorphism, as desired.

19. NUMBER OF ENDOMORPHISMS

Since there are usually lots of endomorphisms in a BA, the variations of this function under algebraic operations have not been studied much. Its main relationships to our other functions are the following two easily established facts: $|\mathrm{Ult}A| \le |\mathrm{End}A|$ and $|\mathrm{Aut}A| \le |\mathrm{End}A|$. If A is the BA of finite and cofinite subsets of an infinite cardinal κ, then $|\mathrm{Ult}A| = \kappa$ while $|\mathrm{Aut}A| = |\mathrm{End}A| = 2^\kappa$. For an infinite rigid BA A we have $|\mathrm{Aut}A| < |\mathrm{End}A|$.

It is more interesting to construct a BA A such that $|A| = |\mathrm{Ult}A| = |\mathrm{End}A|$, and we will spend the rest of this section discussing this. An easy example of this sort is the interval algebra of the reals, and we first want to generalize the argument for this. (Here we are repeating part of Monk [89].)

Theorem 19.1 *Suppose that L is a complete dense linear ordering of power $\lambda \ge \omega$, and D is a dense subset of L of power κ, where $\lambda^\kappa = \lambda$. Let A be the interval algebra on L. Then $|A| = |\mathrm{End}A| = \lambda$.*

PROOF. Recalling the duality for interval algebras from Part I of the BA handbook, we see that Ult$A$ is a linearly ordered space with a dense subset (in the topological sense) of power κ. Therefore there at most $\lambda^\kappa = \lambda$ continuous functions mapping Ult$A$ into itself. The theorem follows. □

Corollary 19.2. *If A is the interval algebra on* **R**, *then $|A| = |\mathrm{End}A| = 2^\omega$.* □

Recalling a construction of more general linear orders of the type described in Theorem 19.1 (see Monk [89]), we get

Corollary 19.3 *If μ is an infinite cardinal and $\forall \nu < \mu(\mu^\nu = \mu)$, then there is a BA A such that $|A| = |\mathrm{End}A| = 2^\omega$.* □

Corollary 19.4 (GCH). *If κ is infinite and regular, then there is a BA A such that $|A| = |\mathrm{End}A| = \kappa^+$.* □

We mention two problems suggested by these results.

Problem 58 (GCH). *If κ is singular or the successor of a singular cardinal, is there a BA A with $|A| = |\mathrm{End}A| = \kappa$?*

Problem 59. *Can one prove in ZFC that there are arbitrarily large cardinals κ for which there is a BA A such that $|A| = |\mathrm{End}A| = \kappa$?*

We mention one more result connecting $|A|$ and $|\mathrm{End}A|$:

Theorem 19.5. *If $|A| = \omega_1$, then $|\mathrm{End}A| \ge 2^\omega$.*

PROOF. If A has an atomless subalgebra, then $|\text{End}A| \geq |\text{Ult}A| \geq 2^\omega$. So suppose that A is superatomic. Then there is a homomorphism f from A onto B, the finite-cofinite algebra on ω: if a is an atom of $A/\langle \text{At}A \rangle^{\text{Id}}$, then f can be taken to be the composition of the natural onto mappings

$$A \to A \restriction a \to C \to B,$$

where C is the finite-cofinite algebra on ω or ω_1. There is an isomorphism g of B into A. If X is any subset of ω with $\omega \backslash X$ infinite, then $B/\langle \{i\} : i \in X \rangle^{\text{Id}}$ is isomorphic to B, and so there is an endomorphism k_X of B with kernel $\langle \{i\} : i \in X \rangle^{\text{Id}}$. Clearly the endomorphisms $g \circ k_X \circ f$ of A are distinct for distinct X's. □

Corollary 19.6 ($\omega_1 < 2^\omega$). *There is no BA A with the property that* $|A| = |\text{End}A| = \omega_1$. □

20. NUMBER OF IDEALS

Again we shall not consider algebraic operations. The main relationships with our earlier functions are: $|\text{Ult}A| \leq |\text{Id}A|$ and $2^{\text{s}A} \leq |\text{Id}A|$; both of these facts are obvious. Also recall the deep Theorem 10.10 from Part I of the BA handbook: if A is an infinite BA, then $|\text{Id}A|^\omega = |\text{Id}A|$. Note that $|\text{Ult}A| < |\text{Id}A|$ for A the finite-cofinite algebra on an infinite cardinal κ.

Next, we show that $|\text{Id}A| = 2^\omega$ for the interval algebra A on the reals; thus A has the property that $|A| = |\text{Ult}A| = |\text{Id}A|$. For each ideal I on A, let $\equiv_I$ be defined as follows: $a \equiv_I b$ iff $a = b$ or else if, say $a < b$, then $[a, b) \in I$. Thus $\equiv_I$ is a convex equivalence relation on $\mathbf{R}$. Now define the function f by setting, for any ideal I,

$$
\begin{aligned}
fI = \{(r, s, \varepsilon) : & \text{ there is an equivalence class } a \text{ under } \equiv_I \text{ such that } |a| > 1 \\
& \text{and } a \text{ has left endpoint } r, \text{ right endpoint } s, \text{ and} \\
& \varepsilon = 0, 1, 2, 3 \text{ according as } a \text{ is } [r, s], [r, s), (r, s], \text{ or } (r, s)\}.
\end{aligned}
$$

Clearly f is a one-one function; since $fI \in (\mathbf{R} \times \mathbf{R} \times 4)^{\leq\omega}$, it follows that $|\text{Id}\mathbf{R}| = 2^\omega$, as desired.

A rigid BA A shows that $|\text{Aut}A| < |\text{Id}A|$ is possible. But the following questions are open.

Problem 60. *Is there a BA A such that* $|\text{End}A| < |\text{Id}A|$*?*

Problem 61. *Is there a BA A such that* $|\text{Id}A| < |\text{Aut}A|$*?*

21. NUMBER OF SUBALGEBRAS

For this last cardinal function, we also do not consider algebraic operations. The main fact connecting it with earlier functions is as follows:

Theorem 21.1. $|\text{Id}A| \leq |\text{Sub}A|$.

PROOF. Let $X = \{I : I \text{ is a proper non-maximal ideal of } A\}$. For all $I \in X$ let $fI = I \cup -I$. Then f is one-one: suppose that $I \neq J$; say that $b \in I \backslash J$. Choose a so that $a, -a \notin J$. Then $a \cdot b \notin J$ or $-a \cdot b \notin J$; say $a \cdot b \notin J$. Then $a \cdot b \in fI \backslash fJ$, as desired. Thus $X \leq |\text{Sub}A|$.

It suffices now to show that $|\text{Ult}A| \leq |X|$. Fix $F \in \text{Ult}A$. For any $G \in \text{Ult}A\backslash\{F\}$, let $gG = F \cap G$. Then g is a one-one function from $\text{Ult}A\backslash\{F\}$ into X: if F, G, H are distinct ultrafilters, choose $x \in G\backslash H$ and $z \in F\backslash H$; then $x + z \in fG\backslash fH$. □

Also note that $2^{\text{Irr}A} \leq |\text{Sub}A|$. An example A for which $|\text{Id}A| < |\text{Sub}A|$ is provided by the interval algebra A on the reals. We noted in the last section that $|\text{Id}A| = 2^\omega$. Since $\text{Irr}A = 2^\omega$, we have $|\text{Sub}A| = 2^{2^\omega}$.

Theorem 21.1 implies that $|\text{Sub}A|$ is our biggest cardinal function. Its size is, of course, always at most $2^{|A|}$. For most algebras, this value is actually attained. It is also quite interesting to construct a BA A in which $|\text{Sub}A|$ is as small as possible. The only algebra we know of where this is the case is Rubin's algebra A from section 16. We now go through the proof that $|\text{Sub}A| = \omega_1$; thus $|A| = |\text{Ult}A| = |\text{Id}A| = |\text{Sub}A|$. We call an element $a \in A$ *countable* provided that $A \upharpoonright a$ is countable.

Lemma 21.2. *A has only countably many countable elements.*

PROOF. Let P be the set of all countable elements of A, and suppose that P is uncountable. Then it is easy to construct $\langle a_\alpha : \alpha < \omega_1 \rangle \in {}^{\omega_1}P$ such that if $\alpha < \beta < \omega_1$ then $a_\beta \not\leq a_\alpha$. Since $\text{h-cof}(A) = \omega$, the set $P' \stackrel{\text{def}}{=} \{a_\alpha : \alpha < \omega_1\}$ is not well-founded; say $a_{\alpha(0)} > a_{\alpha(1)} > \cdots$. Choose $i, j \in \omega$ such that $i < j$ and $\alpha(i) < \alpha(j)$. Then $a_{\alpha(i)} > a_{\alpha(j)}$ contradicts the choice of the a_β's. □

Note that the collection of countable elements of A forms an ideal, which we denote by $C(A)$. To proceed further, we have to go back to the main property of Rubin's algebra. For this purpose we introduce the following notation. Suppose that P is a subset of A, and $a, b_1, \ldots, b_n$ are elements of A. We say that P *is dense at* $\langle a, b_1, \ldots, b_n \rangle$ provided that for all c_1, c_2, if $\langle a_{,1}, c_2, b_1, \ldots, b_n \rangle$ is a configuration then $P \cap (c_1, c_2) \neq 0$. Thus the main property of A is that if P is an uncountable subset of A then there exist $n \in \omega$ and disjoint $a, b_1, \ldots, b_n \in A$ such that P is dense at $\langle a, b_1, \ldots, b_n \rangle$. For both of the next two lemmas we advise the reader to draw a diagram along the lines of the one in section 16 to see what is going on.

Lemma 21.3. *Assume that $P \subseteq A$, P is uncountable, P is dense at $\langle a, b_1, \ldots, b_n \rangle$, $a \leq b \leq a + \sum_{i=1}^n b_i$, $b \cdot b_i \neq 0$ for $i = 1, \ldots, n$, and $b \cdot -a \notin C(A)$. Then $P \cap [a, b]$ is uncountable.*

PROOF. Suppose that $P \cap [a, b]$ is countable, and let Q be the closure of $P \cap [a, b]$ under $+$; so Q is countable also. Say $b \cdot b_i \notin C(A)$. Pick any $c \leq b \cdot b_i$, $c \neq 0$, such that $c' \stackrel{\text{def}}{=} a + c + \sum_{j \neq i} b_j \notin Q$. Then pick c_1, c_2 so that $a + c_i < c'$ and $\langle a, c_i, c', b_1, \ldots, b_n \rangle$ is a configuration for $i = 1, 2$, and $c_1 + c_2 = c'$. Then pick $d_1, d_2 \in P$ so that $c_i < d_i < c'$ for $i = 1, 2$. But $d_1, d_2 \in [a, b]$ and $d_1 + d_2 = c'$, so $c' \in Q$, contradiction. □

Lemma 21.4. *Every subalgebra of A is the union of countably many closed intervals.*

PROOF. Suppose that B is a subalgebra of A which is not the union of countably many closed intervals. Let $\langle [x_\alpha, y_\alpha] : \alpha < \omega_1 \rangle$ enumerate all of the closed intervals contained in B. Now B contains ω_1 elements pairwise inequivalent with respect to $C(A)$; hence it is easy to construct a sequence $\langle z_\alpha : \alpha < \omega_1 \rangle \in {}^{\omega_1}B$ with the following two properties:

(1) $z_\alpha \notin \bigcup_{\beta < \alpha} [x_\beta, y_\beta]$ for each $\alpha < \omega_1$;
(2) $z_\alpha \triangle z_\beta \notin C(A)$ for distinct $\alpha, \beta < \omega_1$.

Let $D = \{z_\alpha : \alpha < \omega_1\}$. Since D is somewhere dense, choose $n \in \omega$ and disjoint $a, b_1, \ldots, b_n$ such that each $b_i \neq 0$ and D is dense at $\langle a, b_1, \ldots, b_n \rangle$. Choose any b such that $a \leq b \leq a + \sum_{i=1}^n b_i$ and $b \cdot b_i \neq 0 \neq b_i \cdot -b$ for all $i = 1, \ldots, n$. We show that $[a, b] \subseteq B$. Take any $d \in [a, b]$. Choose e_1, e_2 with the following properties: $d = e_1 \cdot e_2$; $a \leq e_i \leq a + \sum_{j=1}^n b_j$; $e_i \cdot b_j \neq 0$ for $i = 1, 2$, $j = 1, \ldots, n$. Thus $\langle a, d, e_i, b_1, \ldots, b_n \rangle$ is a configuration, so we can choose $f_i \in D \cap (d, e_i)$ for $i = 1, 2$. Then $f_1 \cdot f_2 = d$, and so $d \in B$ (since $D \subseteq B$). Thus, indeed, $[a, b] \subseteq B$. By an easy argument, $[a, b] \cap D$ has at least two elements. Since distinct elements of D are inequivalent mod $C(A)$, it follows that $b \cdot -a \notin C(A)$. Hence by Lemma 21.3, $D \cap [a, b]$ is uncountable. But this clearly contradicts the construction of D. □

With these lemmas available we can now prove that $|\text{Sub}A| = \omega_1$. We claim that each subalgebra B of A is generated by an ideal along with a countable set. In fact, write $B = \bigcup_{i < \omega} [a_i, b_i]$. Let I be the ideal generated by $\{b_i \cdot -a_i : i < \omega\}$. Then clearly B is generated by $I \cup \{b_i : i < \omega\}$, as required. Now every ideal is countably generated. This follows from the fact that $\text{hL}A \leq \text{h-cof}A = \omega$, proved in section 16, and then one of the equivalents of hL given in section 13 gives the desired assertion. This being the case, it follows that there are exactly ω_1 ideals in A, since $\omega_1^\omega = \omega_1$ by virtue of CH (which follows from $\diamondsuit$, which we are assuming). Now $|\text{Sub}A| = \omega_1$ is clear, again using CH.

Problem 62. *Can one construct in ZFC a BA A with $|A| = |\text{Sub}A|$?*

It is possible to have $|\text{Aut}A| < |\text{Sub}A|$: take a rigid BA. One can even have $|\text{End}A| < |\text{Sub}A|$, for example in the interval algebra on the reals. The following question is open, however.

Problem 63. *Is there a BA A such that* $|\text{Sub}A| < |\text{Aut}A|$*?*

DIAGRAMS
1. GENERAL CASE

$l =$ difference can be large; $s =$ "small" difference

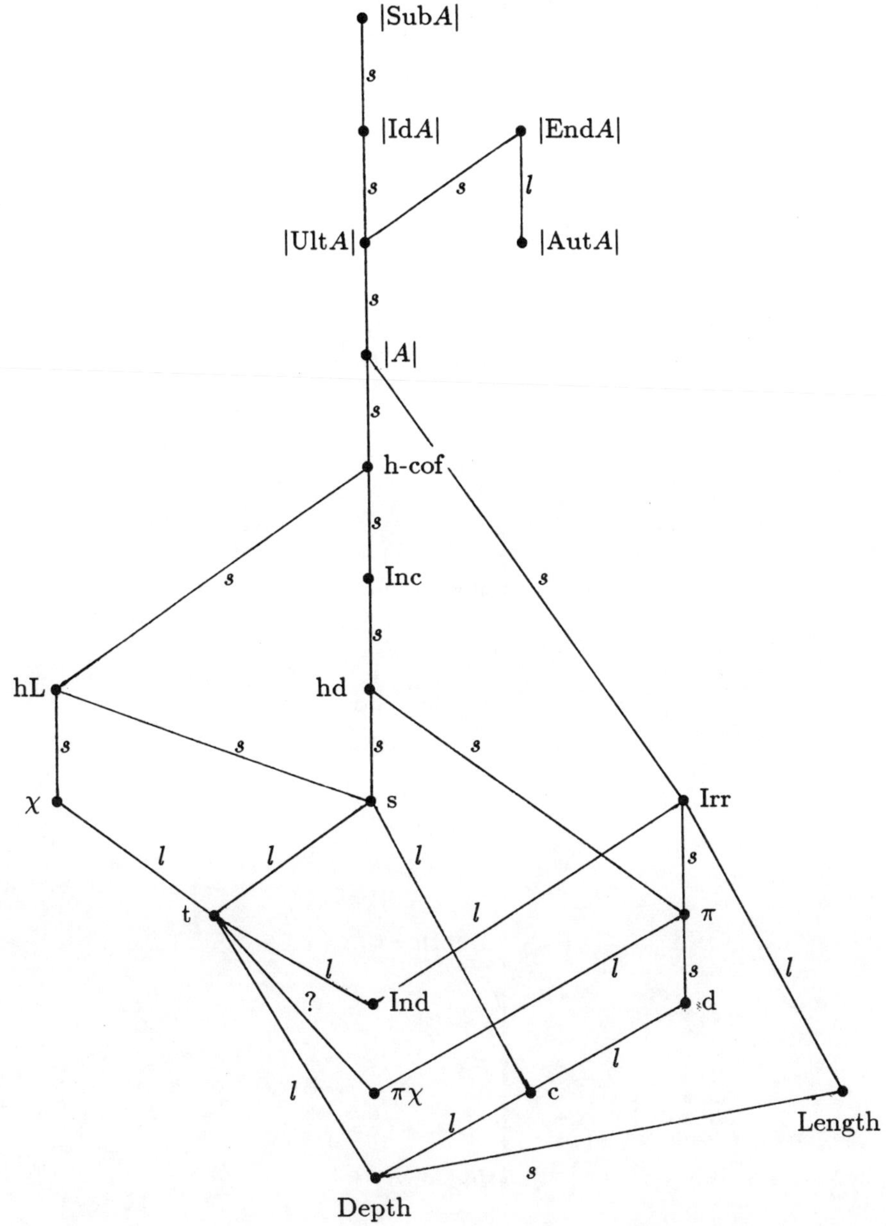

2. INTERVAL ALGEBRAS

$l =$ difference can be large; $s =$"small" difference

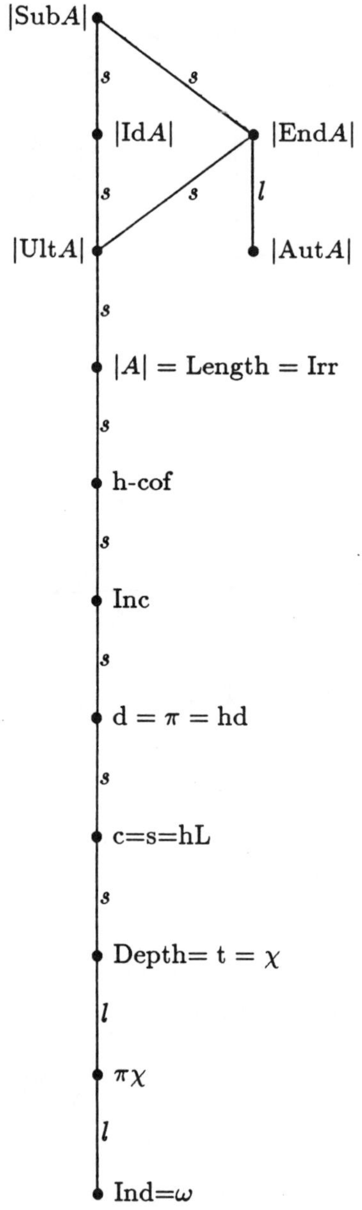

3. COMPLETE BA'S

$l =$ difference can be small; $s =$"small" difference

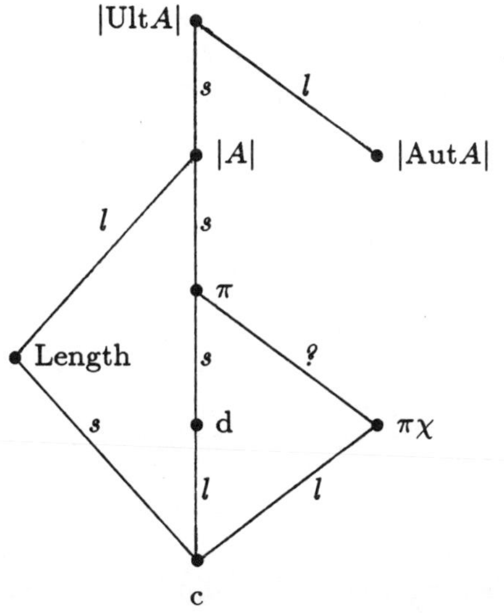

c=Depth
$|A| = \text{Ind} = \text{t} = \text{s} = \chi = \text{hL} = \text{Irr} = \text{s} = \text{hd} = \text{Inc} = \text{h-cof}$
$|\text{Ult}A| = |\text{Id}A| = |\text{Sub}A| = |\text{End}A|$

The above diagrams are supposed to summarize the material of this book. We now give some indications of the theorems or examples involved, with references to the text. For each of the three diagrams, we need to do the following: (1) For each edge, indicate where the relation is proved, and give an example where the functions involved are different. Also, if the difference is indicated as "small", indicate where that is stated in the text, while if the difference is "large", indicate an example. (2) Show that there are not any relations except those indicated in the diagrams. It suffices to do this just for crucial places in the diagram. For example, if in the main diagram we give an algebra A in which Length$A < \pi\chi A$, this will also be an example for Length$A <$ h-cof$A$ and Depth$A < \pi\chi A$. As will be seen, we have not been completely successful in either of these two tasks; there are several open problems left.

A. **The main diagram**. Part 1. The edges.

Depth $\leq$ Length. This relation is obvious from the definitions. The difference is small by the Erdös, Rado theorem; see the end of section 5.

Depth $\leq$ c. Obvious from the definitions. The difference is large in the finite-cofinite algebra on an infinite cardinal κ.

Depth $\leq$ t. This is proved in Theorem 2.11. The difference is big in a free algebra.

$\pi\chi \leq$ t. See Theorem 9.5. They differ in $\mathcal{P}\kappa$, but it is an open question whether the difference is always "small"; see Problem 33.

$\pi\chi \leq \pi$. Obvious from the definitions. The difference is large in a finite-cofinite algebra; see the proof of Theorem 9.1.

c $\leq$ s. This is a consequence of Theorem 1.3. The difference is large in free algebras.

c $\leq$ d. See Theorem 3.1. The difference is large in free algebras.

Length $\leq$ Irr. Obvious from the definitions. The difference is large in free algebras.

Ind $\leq$ t. See the beginning of section 10. The difference can be large in some interval algebras, since Depth $\leq$ t.

Ind $\leq$ Irr. Obvious from the definitions. The difference is large in some interval algebras.

d $\leq \pi$. Obvious from the topological versions of these functions. The difference is small: see the end of section 4.

t $\leq \chi$. Obvious from the definitions. The difference is big in a finite-cofinite algebra on a cardinal κ.

t $\leq$ s. See Theorem 11.5. The difference is big in a finite-cofinite algebra on a cardinal κ.

$\pi \leq$ hd. See Theorem 14.1. The difference is small, since $\pi A \leq \mathrm{hd} A \leq |A| \leq 2^{\pi A}$. The functions differ in $\mathcal{P}\kappa$, for example.

$\pi \leq$ Irr. See Theorem 4.25 of Part I of the BA handbook. Again, the difference is small and they differ in $\mathcal{P}\kappa$.

$\chi \leq$ hL. See page 98. The difference is small, since $|\mathrm{Ult} A| \leq 2^{\chi}$ by Theorem 12.5; they differ on the Alexandroff duplicate of a free algebra; see page 91.

s $\leq$ hL. Obvious from the definitions. The difference is small, since $|A| \leq 2^{\mathrm{s}A}$ for any BA A by Theorem 11.7. They differ in a Kunen line (constructed under CH). Whether there is an example in ZFC is an open question (Problem 45).

s $\leq$ hd. Obvious from the definitions. The difference is small (see above). They differ on the interval algebra of a Souslin line. Whether there is an example in ZFC is open (Problem 50).

Irr $\leq$ Card. Obvious from the definitions. The difference is small; from Theorem 4.23 of Part I it follows that $|A| \leq 2^{\mathrm{Irr} A}$. A compact Kunen line (constructed under CH) gives a BA in which they are different (see section 6). It is open to give an example in ZFC (Problem 21).

hL $\leq$ h-cof. See page 108. The difference is small, since $\chi A \leq \mathrm{hL} A$, and so

$\mathrm{hL}A \leq \text{h-cof}A \leq |A| \leq 2^{\chi A} \leq 2^{\mathrm{hL}A}$, using Theorem 12.5. They differ on the interval algebra on the reals; see the beginning of section 16.

hd $\leq$ Inc. This is an easy consequence of Theorem 4.25 of the BA handbook, part I. The difference is small, since s $\leq$ hd, Inc $\leq$ Card, and $|A| \leq 2^{\mathrm{s}A}$ for any BA (Theorem 11.7).

Inc $\leq$ h-cof. Obvious from the definitions. The difference is small since by the above $|A| \leq 2^{\mathrm{Inc}A}$. They differ on the Baumgartner, Komjath algebra (see the beginning of section 16); this was constructed using $\diamondsuit$, and it remains a problem to get an example with weaker assumptions (Problem 55).

h-cof $\leq$ Card. Obvious from the definitions. The difference is small (see above). An example where they differ is the algebra of Bonnet and Shelah [85].

Card $\leq$ |Ult|. This is well-known. The difference is, of course, small. They differ in an infinite free algebra.

|Ult| $\leq$ |End|. This is obvious. The difference is small. They differ for the finite-cofinite algebra on an infinite cardinal.

|Aut| $\leq$ |End|. Also obvious. The difference is large, as shown by any rigid BA.

|Ult| $\leq$ |Id|. Again obvious. The difference is small. They differ on the finite-cofinite algebra on an infinite cardinal.

|Id| $\leq$ |Sub|. See Theorem 21.1. The difference is small. They differ for the interval algebra on the reals.

B. **The main diagram**. Part 2. No other relationships. Keep in mind that we only treat "crucial" relations; other possibilities are supposed to follow from these.

Length $< \pi\chi$: an uncountable free algebra: see Theorem 9.3.
Length $<$ c: the finite-cofinite algebra on an uncountable cardinal.
Length $<$ Ind: an uncountable free algebra.
$\pi\chi <$ Depth: see the example on page 77.
d $< \pi\chi$: some free algebras.
Ind $<$ Depth: the interval algebra on an uncountable cardinal.
Ind $< \pi\chi$: true in the interval algebra on an uncountable cardinal; see page 77.
$\pi <$ Ind: $\mathcal{P}\kappa$.
$\chi <$ c: the Alexandroff duplicate of an infinite free algebra; see page 93.
Irr $<$ t: no example is known; this is Problem 34. There are weaker problems which were not explicitly stated, e.g., is there an example with Irr $<$ s?
hL $<$ d: The interval algebra of a complete Souslin line. It is not known if this is possible in ZFC (Problem 46)
hL $<$ Inc: The interval algebra on the reals; see page 99.
Inc $< \chi$: The Baumgartner-Komjath algebra, constructed under $\diamondsuit$; see section 15. It is not known if this is possible under weaker hypotheses (Problem 54). Weaker problems are hd $< \chi$? and s $< \chi$? (Problems 51,40).

h-cof < Length: The Bonnet-Shelah algebra.
|Aut| < Depth: embed a large interval algebra in a rigid algebra.
|Aut| < $\pi\chi$: embed a large free algebra in a rigid algebra; see Theorem 9.6.
|Aut| < Ind: embed a large free algebra in a rigid algebra.
|Sub| < |Aut|: an example is not known (Problem 63).
|Id| < |Aut|: an example is not known (Problem 61).
|Ult| < |Aut|: the finite-cofinite algebra on an infinite cardinal.

C. **The interval algebra diagram.** Part 1. The indicated equalities, and the "large" and "small" indications (the edges follow from the general diagram).

Ind = ω: this is one of the main results about interval algebras; see Part I of the BA handbook.
$\omega \le \pi\chi$, difference possibly large. See the description of $\pi\chi$ for interval algebras on page 79.
Depth = t = χ: see page 95.
$\pi\chi$ < Depth, difference possibly large. See the example on page 79 (which has an obvious generalization).
c=s=hL: see page 99.
Depth $\le$ c, with the difference small. The difference is small since $|A| \le 2^{\mathrm{Depth}A}$ for an interval algebra A; *this implies smallness for the next few that we consider also.* For an example where they differ, see page 39.
d = π = hd: This is an obvious consequence of the retractiveness of interval algebras.
c $\le$ d. They differ in the interval algebra of a Souslin line; see page 45.
d $\le$ Inc. They differ in the interval algebra on the reals.
Inc $\le$ h-cof. No interval algebra example where they differ is known.
h-cof $\le$ Card. They differ in the Bonnet-Shelah algebra.
Card $\le$ |Ult|. They differ for the interval algebra on the rationals.
|Ult| $\le$ |End|. They differ on the interval algebra on κ.
|Aut| $\le$ |End|, the difference large: take an infinite rigid interval algebra.
|Ult| $\le$ |Id|. They differ on the interval algebra on κ.
|Id| $\le$ |Sub|. They differ on the interval algebra on the reals.
|End| $\le$ |Sub|. They differ on the interval algebra on the reals.

D. **The interval algebra diagram.** Part 2. No other relationships.

|Aut| < Ind: an infinite rigid interval algebra.
|Id| < |Aut|: no example is known (Problem 61).
|Ult| < |Aut|: the interval algebra on κ.

E. **The complete BA diagram.** Part 1. The indicated equalities and inequalities and the "large" and "small" indications. In fact, the "small" indications are clear.

c = Depth: obvious.
c < Length: $\mathcal{P}\omega$.
c < d: free algebras.
c < $\pi\chi$: free algebras; see the end of section 9, and Theorem 9.3.
d < π: completion of the free algebra on ω_1 free generators.
$\pi\chi < \pi$: no complete BA with this property is known (Problem 27).
Length < Card: a large ccc algebra.
π < Card: $\mathcal{P}\kappa$.
Ind = Card, |Ult| = |Sub|, and Card < |Ult|: all true by the Balcar, Franěk theorem.
|Aut| < |Ult|: a rigid algebra.

F. The complete BA diagram. Part 2. No other relations.
|Aut| < c: embed $\mathcal{P}\kappa$ in a rigid BA.
Card < |Aut|: the completion of the free BA of size 2^ω.

EXAMPLES

We determine, as much as possible, our cardinal functions on the following examples; see also the following table:

1. The finite-cofinite algebra on κ.
2. The free algebra on κ free generators.
3. The interval algebra on the reals.
4. $\mathcal{P}\kappa$.
5. The interval algebra on κ.
6. $\mathcal{P}\omega/\mathrm{Fin}$.
7. The Alexandroff duplicate of a free algebra.
8. The completion of a free algebra.
9. The countable-cocountable algebra on ω_1.
10. A compact Kunen line.
11. The Baumgartner-Komjath algebra.
12. The Rubin algebra.

We do not have to consider all of our 21 functions for each of them, since usually the determination of some key functions says what the rest are.

1. The finite-cofinite algebra on κ.

1. The length is ω, by an easy argument.
2. $\mathrm{c}A = \kappa$.
3. $\mathrm{t}A = \omega$. See section 10.
4. $\chi A = \kappa$. See section 12.
5. $|\mathrm{Ult}A| = \kappa$.
6. $|\mathrm{Aut}A| = 2^\kappa$.
7. $|\mathrm{Id}A| = 2^\kappa$.

2. The free BA on κ free generators.

1. $\mathrm{Length}A = \omega$. See Handbook, Part I, Corollary 9.17.
2. $\mathrm{d}A =$ the smallest cardinal λ such that $\kappa \le 2^\lambda$; see Corollary 3.5.
3. $\pi\chi A = \kappa$: see Theorem 9.3.
4. $\mathrm{Ind}A = \kappa$.
5. $|\mathrm{Ult}A| = 2^\kappa$.
6. $|\mathrm{Aut}A| = 2^\kappa$.

3. The interval algebra on the reals.

1. $\mathrm{hd}A = \omega$. See section 14 and Lemma 1.6.
2. $\mathrm{Inc}A = 2^\omega$. See section 15.
3. $|\mathrm{Id}A| = 2^\omega$. See section 20.
4. $|\mathrm{End}A| = 2^\omega$; Corollary 19.2.

5. $|\text{Aut}A| = 2^\omega$; this is clear by the above, since it is easy to exhibit 2^ω automorphisms.

6. $|\text{Sub}A| = 2^{2^\omega}$. This follows since $|\text{Sub}A| = 2^{|A|}$ for every interval algebra A.

4. $\mathcal{P}\kappa$.

1. $cA = \kappa$.
2. $\pi A = \kappa$.
3. $\text{Length}A = \text{Ded}\kappa$. See Theorem 5.4.
4. $\pi\chi A = \kappa$. In fact, let Φ be the collection of finite subsets of κ, and Ψ the collection of finite subsets of Φ. Thus $|\Phi| = |\Psi| = \kappa$, and we work with $\mathcal{P}(\Phi \times \Psi)$ rather than $\mathcal{P}\kappa$. For each $X \subseteq \kappa$ let $a_X = \{(F,G) : F \in \Phi, G \in \Psi,$ and $X \cap F \in G$. These sets are independent, since if M and N are disjoint finite subsets of $\mathcal{P}\kappa$, choose $F \in \Phi$ such that $X \cap F \neq Y \cap F$ for distinct $X, Y \in M \cup N$, and let $G = \{F \cap X : X \in M\}$; then it is easy to see that $(F,G) \in \bigcap_{X\in M} a_X \cap \bigcap_{X\in N} -a_X$. Now let H be an ultrafilter on $\mathcal{P}(\Phi \times \Psi)$ such that $a_X \in H$ whenever $X \subseteq \kappa$ and $X \neq \kappa$, while $-a_\kappa \in H$. Suppose that $\pi\chi H < \kappa$; let Z be a π-basis for H of size $< \kappa$. We may assume that the members of Z are singletons. Let $X = \bigcup_{\{(F,G)\}\in Z} F$; so, $X \neq \kappa$. Choose $\{(F,G)\} \in Z$ such that $\{(F,G)\} \subseteq a_X \backslash a_\kappa$. Since $X \cap F = F$, we have $F \in G$. But this means that $(F,G) \in a_\kappa$, contradiction.
5. $|\text{Ult}A| = 2^{2^\kappa}$.
6. $|\text{Aut}A| = 2^\kappa$.

5 The interval algebra on κ.

1. $\pi\chi A = \kappa$. See the end of section 9. $\pi\chi A$ is attained if κ is regular, otherwise not.
2. $|\text{Ult}A| = \kappa$. See Theorem 17.10 of Part I of the BA handbook.
3. $|\text{Aut}A| = 2^\kappa$. See the end of section 18.
4. $|\text{Id}A| = 2^\kappa$. This is true since $sA = \kappa$.

6. $\mathcal{P}\omega/\text{Fin}$

1. $\text{Depth}A \geq \omega_1$. It is consistent that it is ω_1 and $\neg$CH holds. Under MA, it is 2^ω. See McKenzie, Monk [82] for references.
2. $\text{Length}A = 2^\omega$.
3. $cA = 2^\omega$.
4. $\pi\chi A \geq \text{cf}2^\omega$. $\text{Con}(2^\omega = \aleph_{\omega_1} + \pi\chi A = \omega_1)$; see van Mill [84], p. 558.
5. $\text{Ind}A = 2^\omega$.
6. $|\text{Ult}A| = 2^{2^\omega}$.
7. $|\text{Aut}A|$ can consistently be 2^ω or 2^{2^ω}; see van Mill [84], p. 537.

7. The Alexandroff duplicate of a free BA.

We use notation as in section 12. Thus B is a free BA of size κ and Dup$B$ is its Alexandroff duplicate.

1. $\mathrm{c}(\mathrm{Dup}B) = 2^\kappa$.

2. $\chi\mathrm{Dup}B = \kappa$. See p. 91.

3. $\mathrm{Length}(\mathrm{Dup}B) = \omega$. In fact, suppose that $Y \subseteq \mathrm{Dup}B$, Y a chain, $|Y| = \omega_1$. Define $(a, X) \equiv (b, Z)$ iff $(a, X), (b, Z) \in Y$ and $a = b$. Then since B has no uncountable chains, there are only countably many $\equiv$-classes. So there is a class, say K, which has ω_1 elements; say that a is the first member of each ordered pair in K. Then $M \stackrel{\text{def}}{=} \{X : (a, X) \in K\}$ is of size ω_1, is a chain under inclusion, and if $X, Z \in M$ with $X \subseteq Z$, then $Z \backslash X$ is finite. Clearly this is impossible.

4. $\mathrm{Ind}(\mathrm{Dup}B) = \kappa$.

5. $\pi\chi\mathrm{Dup}B = \kappa$. For, let F be a non-principal ultrafilter on Dup$B$, and suppose that D is dense in F. Set $G = \{b \in B : (b, X) \in F \text{ for some } X\}$. Then G is an ultrafilter on B, and $\{b : (b, X) \in D \text{ for some } X\}$ is dense in G. Hence $|D| \geq \kappa$. So $\pi\chi\mathrm{Dup}B = \kappa$ since $\pi\chi \leq \chi$.

6. $|\mathrm{Ult}(\mathrm{Dup}(B))| = 2^\kappa$. See the description of ultrafilters in section 12.

7. $|\mathrm{Id}(\mathrm{Dup}(B))| = 2^{2^\kappa}$.

8. We have not been able to determine the size of Aut(Dup$B$)). It is at least 2^κ.

Problem 64. *How many automorphisms does the Alexandroff duplicate of the free algebra on κ generators have?*

8. The completion of a free algebra.

Let B be a free algebra of size κ, A its completion.

1. $\mathrm{c}A = \omega$.

2. $\mathrm{Length}A = 2^\omega$. In fact, $\geq$ is clear. Suppose that L is a chain of size $(2^\omega)^+$. Using a well-ordering of L and the partition relation $(2^\omega)^+ \to (\omega_1)^2_\omega$, we get an uncountable well-ordered chain in A, contradiction.

3. $\mathrm{d}A$ is the least cardinal λ such that $\kappa \leq 2^\lambda$. For, this is true for B itself by Corollary 3.5, and an application of Sikorski's extension theorem shows that it is true of A.

4. $\pi\chi A = \kappa$. For, $\leq$ is clear. Suppose that F is an ultrafilter on A and D is a π-base for F. Without loss of generality $D \subseteq B$. Then D is dense in $F \cap B$, so $|D| \geq \kappa$.

5. $\pi A = \kappa$ by the same argument.

6. $|A| = \kappa^\omega$.

7. $|\mathrm{Ult}A| = 2^{\kappa^\omega}$.

8. $|\mathrm{Aut}A| = 2^\kappa$. In fact, any automorphism of A is uniquely determined by its restriction to B, and there are only 2^κ mappings of B into A. On the other hand, there are at least 2^κ automorphisms of A.

9. The countable-cocountable algebra on ω_1.

1. Depth$A = \omega_1$, by an easy argument.
2. Length$A = 2^\omega$.
3. $\pi A = \omega_1$.
4. Ind$A = 2^\omega$.
5. $\pi\chi A = \omega_1$: let F be the ultrafilter of cocountable sets. Suppose that D is dense in F, with $|D| \leq \omega$. Without loss of generality the members of D are singletons. But then $\omega_1 \backslash D \in F$, contradiction.
6. $|A| = 2^\omega$.
7. $|\text{Ult}A| = 2^{2^\omega}$.
8. $|\text{Aut}A| = 2^{\omega_1}$.

10. A compact Kunen line.

Recall that this Boolean algebra was constructed using $\diamondsuit$.

1. Irr$A = \omega$. See section 6.
2. $\chi A = \omega_1$. This is clear from the construction.
3. $|A| = \omega_1$. Clear from the construction.

Problem 65. *Determine* Inc$A$, $|\text{Ult}A|$, $|\text{Aut}A|$, $|\text{End}A|$, $|\text{Id}A|$, *and* $|\text{Sub}A|$ *for the compact Kunen line of section 6.*

11. The Baumgartner, Komjath algebra.

Recall that this BA was constructed using $\diamondsuit$. See section 15.

1. Inc$A = \omega$.
2. Length$A = \omega$.
3. $\chi A = \omega_1$.

Problem 66. *Determine* Irr$A$, $|\text{Ult}A|$, $|\text{Aut}A|$, $|\text{End}A|$, $|\text{Id}A|$, *and* $|\text{Sub}A|$ *for the Baumgartner, Komjath algebra.*

12. The Rubin algebra.

See section 16. Some of the facts mentioned here are not proved in these notes, but their proofs are similar to things we showed about Rubin's algebra, and can be found in Rubin [83].

1. h-cof$A = \omega$.
2. Irr$A = \omega$.
3. $|A| = \omega_1$.

3. $|\text{Sub}A| = \omega_1$.
4. $|\text{End}A| = \omega_1$.

We do not know about $|\text{Aut}A|$, although the construction can be changed to make A rigid.

TABLE OF EXAMPLES

Example	Depth	$\pi\chi$	c	Length	Ind	d	t	π	χ	s	Irr	hL	hd	Inc
1. Fincoκ	ω	ω	κ	ω	ω	κ	ω	κ	κ	κ	κ	κ	κ	κ
2. Frκ	ω	κ	ω	ω	κ	(1)	κ	κ	κ	κ	κ	κ	κ	κ
3. Intalg$\mathbf{R}$	ω	ω	ω	2^ω	ω	ω	ω	ω	ω	ω	2^ω	ω	ω	2^ω
4. $\mathcal{P}\kappa$	κ	κ	κ	Dedκ	2^κ	κ	2^κ	κ	2^κ	2^κ	2^κ	2^κ	2^κ	2^κ
5. Intalgκ	κ	κ	κ	κ	ω	κ	κ	κ	κ	κ	κ	κ	κ	κ
6. $\mathcal{P}\omega$/Fin	(2)	(3)	2^ω	2^ω	2^ω	2^ω	2^ω	2^ω	2^ω	2^ω	2^ω	2^ω	2^ω	2^ω
7. Dup	ω	κ	2^κ	ω	κ	2^κ	κ	2^κ	κ	2^κ	2^κ	2^κ	2^κ	2^κ
8. $\overline{\text{Fr}\kappa}$	ω	κ	ω	2^ω	κ^ω	(1)	κ^ω	κ	κ^ω	κ^ω	κ^ω	κ^ω	κ^ω	κ^ω
9. Cblcoω_1	ω_1	ω_1	ω_1	2^ω	2^ω	ω_1	2^ω	ω_1	2^ω	2^ω	2^ω	2^ω	2^ω	2^ω
10. CKL	ω	ω	ω	ω	ω	ω	ω	ω	ω_1	ω	ω	ω_1	ω	?
11. BK	ω	ω	ω	ω	ω	ω	ω	ω	ω_1	ω	?	ω_1	ω	ω
12. Rubin	ω	ω	ω	ω	ω	ω	ω	ω	ω	ω	ω	ω	ω	ω

Notes: Dup is the Alexandroff duplicate of the free BA of size κ.
CKL is the compact Kunen line constructed in section 6.
BK is the Baumgartner, Komjath algebra constructed in section 15.
Rubin is the algebra constructed in section 16.

(1) The least λ such that $\kappa \leq 2^\lambda$.
(2) The depth is $\geq \omega_1$; various possibilities are consistent.
(3) $\geq \mathrm{cf}2^\omega$.

(*Table continued on the next page*)

Example	h-cof	Card	|Ult|	|Aut|	|Id|	|End|	|Sub|
1. Fincoκ	κ	κ	κ	2^κ	2^κ	2^κ	2^κ
2. Frκ	κ	κ	2^κ	2^κ	2^κ	2^κ	2^κ
3. Intalg$\mathbf{R}$	2^ω	2^ω	2^ω	2^ω	2^ω	2^ω	2^ω
4. $\mathcal{P}\kappa$	2^κ	2^κ	2^{2^κ}	2^{2^κ}	2^{2^κ}	2^{2^κ}	2^{2^κ}
5. Intalgκ	κ	κ	κ	2^κ	2^κ	2^κ	2^κ
6. $\mathcal{P}\omega/\mathrm{Fin}$	2^ω	2^ω	2^{2^ω}	(1)	2^{2^ω}	2^{2^ω}	2^{2^ω}
7. Dup	2^κ	2^κ	2^κ	?	2^{2^κ}	?	2^{2^κ}
8. $\overline{\mathrm{Fr}\kappa}$	κ^ω	κ^ω	2^{κ^ω}	2^κ	2^{κ^ω}	2^{κ^ω}	2^{κ^ω}
9. Cblcoω_1	2^ω	2^ω	2^{2^ω}	2^{ω_1}	2^{2^ω}	2^{2^ω}	2^{2^ω}
10. CKL	ω_1	ω_1	?	?	?	?	?
11. BK	ω_1	ω_1	?	?	?	?	?
12. Rubin	ω	ω_1	ω_1	?	ω_1	ω_1	ω_1

Notes: Dup is the Alexandroff duplicate of the free BA of size κ.
CKL is the compact Kunen line constructed in section 6.
BK is the Baumgartner, Komjath algebra constructed in section 15.
Rubin is the algebra constructed in section 16.

(1) Consistently 2^ω or 2^{2^ω}.

REFERENCES

Arhangelskiĭ, A. [78]. *The structure and classification of topological spaces and cardinal invariants.* Russian Mathematical Surveys 33, no. 6, 33-96.

Baumgartner, J. [76]. *Almost disjoint sets, the dense set problem and the partition calculus*, Ann. Math. Logic 9, 401-439.

Baumgartner, J. [80]. *Chains and antichains in $\mathcal{P}\omega$.* J. Symb. Logic 45, 85-92.

Baumgartner, J., Komjath, P. [81]. *Boolean algebras in which every chain and antichain is countable.* Fund. Math. 111, 125-133.

Baumgartner, J., Taylor, A., Wagon, S. [82]. **Structural properties of ideals.** Dissert. Math. 197, 95pp.

Bell, M., Ginsburg, J., Todorčević, S. [82]. *Countable spread of* $\exp Y$ *and* λY. Topol. Appl. 14, 1-12.

Bonnet, R., Shelah, S. [85]. *Narrow Boolean algebras.* Annals Pure Appl. Logic 28,1-12.

Burris, S. [75]. *Boolean powers.* Alg. Univ. 5, no. 3, 341-360.

Burris, S., Sankappanavar, H. [81]. **A course in universal algebra.** Springer-Verlag, xvi+276pp.

Chang, C. C., Keisler, H. J. [73]. **Model Theory**. North-Holland, 550pp.

Comfort, W. W. [71]. *A survey of cardinal invariants.* Gen. Topol. Appl. 1, 163-199.

Comfort, W. W., Negrepontis, S. [74]. **The theory of ultrafilters.** Springer-Verlag, x+482 pp.

Comfort, W. W., Negrepontis, S. [82]. **Chain conditions in topology.** Cambridge University Press, xi+300pp.

Cramer, T. [74] *Extensions of free Boolean algebras.* J. London Math. Soc. 8, 226-230.

van Douwen, E. K. [89]. *Cardinal functions on Boolean spaces.* In **Handbook of Boolean algebras**, North-Holland, 417-467.

Erdös, P., Hajnal, A., Máté, A., Rado, R. [84]. **Combinatorial set theory: partition relations for cardinals.** Adadémiai Kiadó, 347pp.

Fedorchuk, V. [75] *On the cardinality of hereditarily separable compact Hausdorff spaces.* Dokl. Akad. Nauk SSSR 222, no. 2, 651-655.

Foreman, M., Laver, R. [88] *Some downwards transfer properties for* $\aleph_2$. Adv. in Math. 67, no. 2, 230-238.

Grätzer, G., Lakser, H. [69] *Chain conditions in the distributive free product of lattices.* Trans. Amer. Math. Soc. 144, 301-312.

Hajnal, A., Juhász, I. [71] *A consequence of Martin's axiom.* Indag. Math. 33, 457-463.

Heindorf, L. [87] *A decidability proof for the theory of countable Boolean algebras in the language with quantification over ideals.* 5th Easter Conference on Model Theory, Sem. Berichte 93, Humboldt Univ., Sekt. Math., 34-45.

Heindorf, L. [89] *A note on irredundant sets.* Alg. Univ. 26, 216-221.

Hodel, R. [84]. *Cardinal functions I.* In **Handbook of set-theoretic topology**, North-Holland, 1-61.

Juhász, I. [71]. **Cardinal functions in topology**. Math. Centre Tracts 34, Amsterdam, 150pp. (This book contains some material not found in its revised version, which follows:)

Juhász, I. [80]. **Cardinal functions in topology - ten years later**. Math. Centre Tracts 123, Amsterdam, 160pp.

Juhász, I. [84]. *Cardinal functions II.* In **Handbook of set-theoretic topology**, North-Holland, 63-109.

Juhász, I., Kunen, K, Rudin, M. E. [76] *Two more hereditarily separable non-Lindelöf spaces*, Can. J. Math 28, 998-1005.

Just, W. [88] *Remark on the altitude of Boolean algebras.* Alg. Univ. 25, 283-289.

Just, W., Koszmider, P. [87] *Remarks on cofinalities and homomorphism types of Boolean algebras.* Preprint.

Keisler, H. J., Prikry, K. [74] *A result concerning cardinalities of ultraproducts.* J. Symb. Logic 39, no. 1, 43-48.

Koppelberg, S. [77] *Boolean algebras as unions of chains of subalgebras*, Alg. Univ. 7.

Kunen, K. [78] *Saturated ideals.* J. Symb. Logic 43, 65-76.

Kunen, K. [80] **Set Theory**, North-Holland, xvi+3l3pp.

Kuratowski, K. [58] **Topologie**, vol. 1, fourth edition, 494pp.

Kurepa, G. [35] *Ensembles linéaires et une classe de tableaux ramifiés (tableaux ramifiés de M. Aronszajn)* Putl. Math. Univ. Belgrade 6, 129-160.

Kurepa, G. [57] *Partitive sets and ordered chains.* "Rad" de l'Acad. Yougoslave 302, 197-235.

Kurepa, G. [62] *The Cartesian multiplication and the cellularity number.* Publ. Inst. Math. (Beograd) (N.S.) 2 (16), 121-139.

Malykhin, V. [72] *On tightness and Souslin number in* exp$X$ *and in a product of spaces*, Sov. Math. Dok. 13, 496-499.

McKenzie, R., Monk, J. D. [82] *Chains in Boolean algebras.* Ann. Math. Logic 22, 137-175.

van Mill, J. *An introduction to* $\beta\omega$. In **Handbook of set-theoretic topology**, North-Holland, 503-567.

Milner, E., Pouzet, M. [86] *On the width of ordered sets and Boolean algebras.* Algebra Universalis 23, 242-253.

Monk, J. D. [83] *Independence in Boolean algebras.* Per. Math. Hung. 14, 269-308.

Monk, J. D. [84]. *Cardinal functions on Boolean algebras.* In Orders: Descriptions and Roles, Annals of Discrete Mathematics 23, 9-37.

Monk, J. D. [89] *Endomorphisms of Boolean algebras.* In **Handbook of Boolean algebras**, North-Holland, 491-516.

Rubin, M. [83] *A Boolean algebra with few subalgebras, interval Boolean algebras, and retractiveness.* Trans. Amer. Math. Soc. 278, 65-89.

Shelah, S. [79] *Boolean algebras with few endomorphisms.* Proc. Amer. Math. Soc. 74, 135-142.

Shelah, S. [80] *Remarks on Boolean algebras.* Alg. Univ. 11, 77-89.

Shelah, S. [83] *Constructions of many complicated uncountable structures and Boolean algebras.* Israel J. Math. 45, 100-146.

Shelah, S. [86] *Strong negative partitions and non productivity of the chain condition.* Abstracts Amer. Math. Soc. 7, no. 4, p. 276.

Shelah, S. [87] Handwritten notes.

Shelah, S. [88a] *Successors of singulars, cofinalities of reduced products of cardinals and productivity of chain conditions.* Israel J. Math. 62, no. 2, 213-256.

Shelah, S. [88b] *On successors of singulars.* Abstracts Amer. Math. Soc. 9, no. 6, p. 500. (no. 88T-03-242)

Shelah, S. [88c] *Jónsson algebras in* $\aleph_{\omega+1}$, preprint.

Shelah, S. [88d] *Strong negative partition relations below the continuum.* To appear, Acta Math. Hungar.

Shelah, S. [89] *Notes in set theory.* Abstracts Amer. Math. Soc. 10, no. 4, 89T-03-125, p. 302.

Shelah, S., Soukup, L. [89] *Some remarks on a question of J. D. Monk.* Preprint.

Solovay, R., Tennenbaum, S. [71] *Iterated Cohen extensions and Souslin's problem.* Ann. of Math. 94, 201-245.

Todorčević, S. [83] *Forcing positive partition relations*, Trans. Amer. Math. Soc. 280, 703-720.

Todorčević, S. [85] *Remarks on chain conditions in products.* Compos. Math. 55, no. 3, 295-302.

Todorčević, S. [86] *Remarks on cellularity in products.* Compos. Math. 57, 357-372.

Todorčević, S. [87] *On the cellularity of Boolean algebras.* Handwritten notes.

Todorčević, S. [89] **Partition problems in general topology.** Contempory Mathematics, v. 84, Amer. Math. Soc., xii+116pp.

Weese, M. [80] *A new product for Boolean algebras and a conjecture of Feiner.* Wiss. Z. Humboldt-Univ. Berlin Math.-Natur. Reihe 29, 441-443.

INDEX OF PROBLEMS

Problem 1. *If κ is strong limit singular,* $\mathrm{c}A = \kappa$, *and* $\mathrm{cf}\kappa \leq \mathrm{c}B < \kappa$, *is* $\mathrm{c}(A \oplus B) = \kappa$*?* Page 16.

Problem 2. *Is there a BA A with* $\mathrm{c}_{\mathrm{Sr}}A = \{(\omega,\omega), (\omega,\omega_1), (\omega_1,\omega_1), (\omega,\omega_2), (\omega_2,\omega_2)\}$*? Equivalently, is there a BA A such that* $|A| = \omega_2 = \mathrm{c}A$*, A has a ccc subalgebra of power ω_2, and every subalgebra of A of size ω_2 either has cellularity ω or ω_2?* Page 30.

Problem 3. Is there a BA with c_{Hr} relation $\{(\omega,\omega),(\omega,\omega_1),(\omega_1,\omega_2)\}$? Page 30.

Problem 4. *Is the following conjecture true? If I is an infinite set, F is a $|I|$-regular ultrafilter on I, and A_i is an infinite BA for every $i \in I$, then in $\prod_{i\in I} A_i/F$ there is a chain of order type $2^{|I|}$.* Page 35.

Problem 5. *Is* $\mathrm{t}B \in \mathrm{Depth}_{\mathrm{Hs}}B$ *for every infinite BA B?* Page 39.

Problem 6. *Is there an infinite cardinal κ and a BA A with the property that* $(\kappa,(2^\kappa)^+) \in \mathrm{Depth}_{\mathrm{Sr}}A$*, while* $(\omega,(2^\kappa)^+) \notin \mathrm{Depth}_{\mathrm{Sr}}A$*?* Page 39.

Problem 7. *Does there exist a BA A such that* $(\omega_1,\omega_2) \in \mathrm{Depth}_{\mathrm{Sr}}A$ *but* $(\omega,\omega_1) \notin \mathrm{Depth}_{\mathrm{Sr}}A$*?* Page 39.

Problem 8. *Is there an infinite cardinal κ and a BA A such that* $(\kappa,(2^\kappa)^+) \in \mathrm{Depth}_{\mathrm{Hr}}A$*, while* $(\omega,(2^\kappa)^+) \notin \mathrm{Depth}_{\mathrm{Hr}}A$*?* Page 40.

Problem 9. *Does there exist a system $\langle A_i : i \in I\rangle$ of infinite BA's and an ultrafilter F on I such that* $d(\prod_{i\in I} A_i/F) < |\prod_{i\in I} \mathrm{d}A_i/F|$*?* Page 42.

Problem 10. *Is it true that always* $\mathrm{d}_{\mathrm{Hs}}A = [\omega,\mathrm{hd}A)$ *or* $\mathrm{d}_{\mathrm{Hs}}A = [\omega,\mathrm{hd}A]$ *for an infinite BA A?* Page 44.

Problem 11. *Is it true that always* $\mathrm{d}_{\mathrm{Ss}}A = [\omega,\mathrm{d}A]$ *for an infinite BA A?* Page 45.

Problem 12. *Is it true that always* $\pi(\prod_{i\in I} A_i/F) = |\prod_{i\in I} \pi A_i/F|$*?* Page 47.

Problem 13. *Is* $\pi_{\mathrm{S}+}A$ *always attained?* Page 51.

Problem 14. *Is it true that for every infinite BA A we have*

$$\pi_{\mathrm{Hs}}A = \begin{cases} [\omega,\mathrm{hd}A], & \textit{if hdA is attained,} \\ [\omega,\mathrm{hd}A), & \textit{otherwise?} \end{cases}$$

Page 52.

Problem 15. *Is it true that for every infinite BA A we have*

$$\pi_{\mathrm{Ss}}A = \begin{cases} [\omega,\pi_{\mathrm{S}+}A], & \textit{if } \pi_{\mathrm{S}+}A \textit{ is attained,} \\ [\omega,\pi_{\mathrm{S}+}A), & \textit{otherwise?} \end{cases}$$

Page 52.

Problem 16. *Let A be the interval algebra of a Souslin line, B a BA with length* $\aleph_{\omega_1}$ *not attained. Is* $\mathrm{Length}(A \oplus B)$ *attained?* Page 55.

Problem 17. *Is it consistent to have a BA A such that* $\omega < \mathrm{Length}A < |A|$, *while A has no homomorphic image of power* $< |A|$*?* Page 56.

Problem 18. *Is always* $\mathrm{Length}_{\mathrm{H-}}A = \mathrm{Card}_{\mathrm{H-}}A$? Page 56.

Problem 19. *Is always* $\mathrm{Length}_{\mathrm{h-}}A = \omega$? Page 56.

Problem 20. *Is always* ${}_{\mathrm{d}}\mathrm{Length}_{\mathrm{S-}}A = \omega$? Page 56.

Problem 21. *Can one construct in ZFC a BA A such that* $\mathrm{Irr}A < |A|$*?* Page 65.

Problem 22. *Is it consistent with CH that there is a BA A such that* $\mathrm{Card}_{\mathrm{Hs}}A = \{\omega, \omega_2\}$? Page 68.

Problem 23. *If* A_i *is a non-trivial interval algebra for each* $i \in I$, *where I is infinite, is* $\mathrm{Ind}(\prod_{i\in I} A_i) = 2^{|I|}$*? Equivalently, is it true that for every infinite cardinal* κ *there is no linear order L and sequence* $\langle x_\alpha : \alpha < (2^\kappa)^+\rangle$ *with the following properties?*

(1) For all $\alpha < (2^\kappa)^+$ *and* $\beta < \kappa$, $x_\alpha\beta$ *is a finite collection of half-open intervals of L.*

(2) For all finite disjoint $\Gamma, \Delta \subseteq (2^\kappa)^+$ *there is a* $\beta < \kappa$ *such that*

$$\bigcap_{\alpha\in\Gamma}(\bigcup x_\alpha\beta) \cap \bigcap_{\alpha\in\Delta}(L\backslash\bigcup x_\alpha\beta) \neq 0.$$

Page 71.

Problem 24. *Can one construct in ZFC a BA A with the property that* $\mathrm{Ind}_{\mathrm{H-}}A < \mathrm{Card}_{\mathrm{H-}}A$? Page 74.

Problem 25. *Is* $\mathrm{Ind}_{\mathrm{h+}}A = \mathrm{Card}_{\mathrm{h+}}A$ *for every infinite BA A*? Page 74.

Problem 26. *Is* $\mathrm{Ind}_{\mathrm{h-}}A = \mathrm{Ind}_{\mathrm{H-}}A$ *for every infinite BA A?* Page 74.

Problem 27. *Is* $\pi A = \pi\chi A$ *for A complete*? Page 81.

Problem 28. *Does attainment of tightness imply attainment in the free sequence sense*? Page 84.

Problem 29. *For* $\mathrm{cf}\kappa > \omega$, *does attainment of tightness in the free sequence sense imply attainment of tightness in the defined sense?* Page 84.

Problem 30 *Does attainment of tightness in the free sequence sense imply attainment of* $\pi\chi_{\mathrm{H+}}$ *?* Page 84.

Problem 31. *Do there exist cardinals* κ *and* λ *such that* λ *is regular,* κ *is singular,* $\mathrm{cf}\kappa < \lambda$, *and there is a sequence* $\langle A_\alpha : \alpha < \kappa\rangle$ *of BA's each with tightness less than* κ *such that* $\bigcup_{\alpha<\lambda} A_\alpha$ *has tightness* κ*?* Page 87.

Problem 32. *Is* $\mathrm{t_{H-}}A = \mathrm{Card_{H-}}A$? Page 87.

Problem 33. *Can the difference between* $\pi\chi$ *and* t *be large*? Page 87.

Problem 34. *Is it possible to have* $\mathrm{t}A > \mathrm{Irr}A$? Page 88.

Problem 35. $\mathrm{s}(\prod_{i\in I} A_i) = \max(2^{|I|}, \sup_{i\in I}\mathrm{s}A_i)$? Page 90.

Problem 36. *Is* $\mathrm{s_{H-}}A = \mathrm{Card_{H-}}A$ *for every infinite BA A*? Page 92.

Problem 37. *Can one prove without GCH that if* χA *is an uncountable regular limit cardinal then* χA *is attained*? Page 94.

Problem 38. *Is* $\chi(\prod_{i\in I} A_i) = \max(2^{|I|}, \sup_{i\in I}\chi A_i)$? Page 94.

Problem 39. *Is* $\chi_{\mathrm{H-}}A = \mathrm{Card_{H-}}A$ *for any infinite BA A*? Page 95.

Problem 40. *Can one construct in ZFC a BA A such that* $\mathrm{s}A < \chi A$*?* Page 96.

Problem 41. *Does attainment of* hL *imply attainment in the right-separated sense?* Page 100.

Problem 42. *Does attainment of* hL *in the right-separated sense imply attainment in the defined sense?* Page 100.

Problem 43. *Does attainment of* hL *in the right separated sense imply attainment in the sense of generation of ideals?* Page 100.

Problem 44. *Is* $\mathrm{hL}(\prod_{i\in I} A_i) = \max(2^{|I|}, \sup_{i\in I}\mathrm{hL}A_i)$ Page 101.

Problem 45. *Is there an example in ZFC of a BA A such that* $\mathrm{s}A < \mathrm{hL}A$*?* Page 102.

Problem 46. *Is there an example in ZFC of a BA A such that* $\mathrm{hL}A < \mathrm{d}A$*?* Page 102.

Problem 47. *Does attainment of* hd *imply attainment in the* $\pi_{\mathrm{H+}}$ *sense?* Page 102.

Problem 48. *Does attainment of* $\pi_{\mathrm{H+}}$ *imply attainment of* hd *in the left-separated sense?* Page 104.

Problem 49. $\mathrm{hd}(\prod_{i\in I} A_i) = \max(2^{|I|}, \sup_{i\in I}\mathrm{hd}A_i)$? Page 104.

Problem 50. *Can one construct in ZFC a BA A such that* $\mathrm{s}A < \mathrm{hd}A$*?* Page 104.

Problem 51. *Can one construct in ZFC a BA A such that* $\mathrm{hd}A < \chi A$*?* Page 104.

Problem 52. *If* $\mathrm{Inc}A$ *is attained in the tree sense, is it also attained in the defined sense?* Page 105.

Problem 53. *Is there an uncountable BA A with countable incomparability and* $|B| > \omega$ *for every infinite homomorphic image B of A*? Page 106.

Problem 54. *Can one construct in ZFC a BA A such that* Inc$A < \chi A$*?* page 110.

Problem 55. *Can one construct in ZFC a BA A such that* Inc$A <$ h-cof$A$*?* Page 111.

Problem 56. *Is there an infinite BA A such that* $|A|$ *and* $|\text{Aut}A|$ *are both smaller than* $|\text{Aut}(A \times A)|$*?* Page 119.

Problem 57. *Is there an infinite BA A such that* $|A|$ *and* $|\text{Aut}A|$ *are both smaller than* $|\text{Aut}(A \oplus A)|$*?* Page 119.

Problem 58 (GCH). *If* κ *is singular or the successor of a singular cardinal, is there a BA A with* $|A| = |\text{End}A| = \kappa$*?* Page 120.

Problem 59. *Can one prove in ZFC that there are arbitrarily large cardinals* κ *for which there is a BA A such that* $|A| = |\text{End}A| = \kappa$*?* Page 120.

Problem 60. *Is there a BA A such that* $|\text{End}A| < |\text{Id}A|$*?* Page 122.

Problem 61. *Is there a BA A such that* $|\text{Id}A| < \text{Aut}A|$*?* Page 122.

Problem 62. *Can one construct in ZFC a BA A with* $|A| = |\text{Sub}A|$*?* Page 124.

Problem 63. *Is there a BA A such that* $|\text{Sub}A| < |\text{Aut}A|$*?* Page 125.

Problem 64. *How many automorphisms does the Alexandroff duplicate of the free algebra on* κ *generators have?* Page 136.

Problem 65. *Determine* Inc$A$, $|\text{Ult}A|$, $|\text{Aut}A|$, $|\text{End}A|$, $|\text{Id}A|$, *and* $|\text{Sub}A|$ *for the compact Kunen line of section 6.* Page 137.

Problem 66. *Determine* Irr$A$, $|\text{Ult}A|$, $|\text{Aut}A|$, $|\text{End}A|$, $|\text{Id}A|$, *and* $|\text{Sub}A|$ *for the Baumgartner, Komjath algebra.* Page 137.

INDEX OF NAMES

Alexandroff, P. S., 4, 20, 67, 94, 96, 102, 130f, 134, 136, 138f, 148
Arhangelskiĭ, A., 1, 36, 96f, 140
Balcar, B., 66, 74, 133
Baumgartner, J., 31, 106, 108, 111, 131, 134, 137ff, 140, 148
Bekkali, M., 110
Bell, M., 57, 140
Bonnet, R., 106, 111, 131f, 140
Burris, S., 9, 11, 140
Carlson, T., 71
Čech, E., 96
Chang, C. C., 14, 140
Comfort, W. W., 1, 14, 29, 35, 140
Cramer, T., 69, 140
van Douwen, E. K., v, 1, 48, 66, 140
Erdös, P., 9, 15, 18, 40, 56, 67, 72, 129, 140
Fedorchuk, V., 30, 140
Foreman, M., 29, 140
Franěk, F., 66, 74, 133
Ginsburg, J., 57, 140
Grätzer, G., 32, 55, 141
Hajnal, A., 15, 52, 140f
Heindorf, L., v, 7, 57, 60, 141
Hodel, R., 1, 141
Hodges, W., 35
Juhász, I., 1, 22, 45, 52, 57, 89, 93, 100, 102, 141
Just, W., 30, 66, 141
Keisler, H. J., 14f, 140f
Komjath, P., 106, 108, 111, 131, 134, 137ff, 148
Koppelberg, S., v, 14, 30, 38, 73f, 141
Koszmider, P., v, 30, 141
Kunen, K., 29, 57, 61, 67, 96, 101, 104, 130, 134, 137ff, 141, 148
Kuratowski, K., 67, 141
Kurepa, G., 16, 45, 54, 67f, 141f
Lakser, H., 32, 55, 141
Laver, R., 16, 29, 140
Malyhin, V., 85, 142
McKenzie, R., 32, 36, 55, 135, 142
Milner, E., 105, 142
Monk, J. D., v, 1, 32, 36, 50, 55, 69, 120, 135, 142f
Negrepontis, S., 14, 29, 35, 140
Nyikos, P., v
Pospıšil, B., 96
Pouzet, M., 105, 142
Prikry, K., 15, 141
Quackenbush, R., 11
Rado, R., 9, 15, 18, 40, 56, 67, 72, 129, 140
Rubin, M., 57, 106, 111ff, 123, 134, 137ff, 142
Rudin, M. E., 57, 141
Sankappanavar, H. P., 9, 140
Shapirovskiĭ, B., 49, 79
Shelah, S., v, 14f, 38, 44f, 54, 68, 71, 105f, 110f, 131f, 140, 142f
Solovay, R., 30, 143
Soukup, L. 71, 143
Tarski, A., 7
Taylor, A., 31, 140
Tennenbaum, S., 30, 143
Todorčević, S., v, 16, 22, 24, 29f, 40, 57, 60, 93, 105, 140, 143
Wagon, S., 31, 140

INDEX OF SYMBOLS

$A[B]$ **11**
$A[B]^*$ **12**
$[\![f = g]\!]$ **9**
$\prod_{B;i\in I} A_i$ **7**
Aut$A$ **118ff**
$\overline{B}$ **7**
cA **7ff**
Card$A$ **66ff**
χA **94ff**
dA **41ff**
Dedκ 54
Depth$A$ **32ff**
End$A$ **120ff**
GsS **8**
h-cof$A$ **111ff**
hd$A$ **103f**
hLA **99ff**
IdA **122f**
Inc$A$ **105ff**
Ind$A$ **69ff**
Irr$A$ **57ff**
k' **4**
$k_{\mathrm{Hs}}A$ **5**
$k_{\mathrm{Ss}}A$ **5**
$k_{\mathrm{Sr}}A$ **6**
$k_{\mathrm{Hr}}A$ **6**
$k_{\mathrm{H}+}A$ **4**
$k_{\mathrm{H}-}A$ **4**
$k_{\mathrm{S}+}A$ **4**
$k_{\mathrm{S}-}A$ **4**
$k_{\mathrm{h}+}A$ **4**
$k_{\mathrm{h}-}A$ **4**
${}_{\mathrm{d}}k_{\mathrm{S}+}A$ **4**
${}_{\mathrm{d}}k_{\mathrm{S}-}A$ **4**
$l_{\mathrm{sup}}A$ **5**
$l_{\mathrm{inf}}A$ **5**
LX **3**
Length$A$ **53ff**
$\mathcal{P}\omega$ 19, passim
$\mathcal{P}\omega$/fin 41, passim
$\mathcal{P}\kappa/I$ 31
πA **47**
$\pi\chi A$ **76ff**
sA **89ff**
Sub$A$ bf 123ff
tA **82ff**
Ult$A$ **117**, passim

INDEX OF WORDS

Alexandroff duplicate 20, 94, 102, 131, 134, 136, 138f
amalgamated free product 4, 17
atomic 6, 24f, 30, 52, 69, 70f, 74f, 79, 85, 89, 94, 117, 121
atomless 6, 38, 106ff, 112f, 115f, 121
attainment 4, 7f, 16, 21f, 28f, 32ff, 38f, 48, 50ff, 69, 71, 76, 79, 82ff, 89f, 94, 100, 103ff, 119, 123, 135, 145ff
automorphism vii, 3, 3, 106, **118ff**, 135f, 148
Balcar-Franěk theorem 66, 74, 133
Baumgartner-Komjath algebra 106, 108, 111, 131, 134, 137ff, 148
Boolean power 4, 11f, 140
Boolean product 4, 9, 11
Borel set 67
caliber 5, 29, 74
ccc 7, 14, 16f, 21, 30, 68, 119, 133, 145
cellularity vii, 1f, 4ff, **7ff**, 34, 38ff, 55, 72ff, 87, 117, 142ff
centered 41
CH 16f, 29f, 39, 44, 52, 57, 60, 67f, 93ff, 101f, 120, 124, 130, 135
chain 4, 7ff, 32ff, 44, 54ff, 67, 73, 84, 87, 95, 109f, 118, 136, 140ff
chain condition 7ff, 140ff
character vii, 2ff, 29, 74, 76ff, **94ff**, 101f, 109, 119
complete subalgebra 31, 81
completion 7, 11, 21, 46, 52, 102, 133ff
configuration 111ff
countably complete 14, 34, 72
CSP 6, 66, 68, 73f
cylindrification 31
Δ-system 17, 18, 65
dense 2, 4, 7, 17, 22ff, 31, 38ff, 54, 59, 63, 65, 74, 76ff, 80f, 87, 92, 103ff, 112ff, 120, 123f, 136ff
density vii, 2f, **41ff**, 48, 104, 107
depth vii, 1ff, **32ff**, 53, 55f, 74, 80f, 87, 93, 98, 110, 119, 127ff, 145
direct product 4, 7ff, 14, 119
discrete 3f, 20f, 24f, 38, 42, 56, 58, 63, 65f, 79f, 80, 89, 91f, 95, 142
endomorphism vii, 3,**120ff**, 142
Erdös, P. 7, 9, 15, 18, 40, 56, 67, 72, 129, 140
finite-cofinite 8, 18, 29f, 35, 39f, 45, 56, 70, 74, 80, 87f, 93ff, 120ff, 130ff
finite intersection property 17, 20, 24, 29f, 37f, 41, 79, 96f, 101
free caliber 74
free extension 112f
free product 4, 11, 15ff, 30, 36, 42, 55, 73, 77f, 85, 90, 101, 104, 119, 141
free sequence 36, 38, 79, 82ff, 87, 91, 98, 146
GCH 16, 29, 39, 44, 94, 120, 147f
generic 107ff
global section 4, 8ff,
Handbook of Boolean algebras v, 1, 7ff, 22, 29, 39f, 57, 68ff, 73f, 105f, 110, 117ff, 120, 122, 130ff, 134f, 140ff
hereditarily separable 60, 140f
hereditary Lindelöf degree vii, 3, **99ff**
hereditary density vii, 3, 48, **103ff**
hereditary cofinality vii, 3, **111ff**
homogeneous 49, 105, 118f
homomorphic spectrum 5, 21

ideal vii, **3**, 8f, 14, 20f, 24, 31, 37, 50, 60, 70, 83f, 89f, 99ff, 106ff, 113, **122ff**, 140f, 147
ideal independence 20ff, 50, 60, 90
incompar vii, 3, **105ff**, 119, 147
independence vii, 2, 5, 14, 16, 20f, 24, 29, 38, 45, 47, 50, 60, 68, **69ff**, 84, 89f, 110, 119, 135, 142
interval algebra 6, 22, 24, 29f, 39f, 46, 50, 52, 55, 66f, 71f, 74f, 77, 80f, 85, 90, 93, 96, 98, 102, 104, 106, 110f, 119ff, 128, 130ff, 134f, 146
irred vii, 2, **57ff**, 141
Kunen line 57, 96, 101, 104, 130, 134, 137f, 148
Kurepa family 67f
length vii, 2, 4f, 22, 32, 35f,**53ff**, 71f, 79f, 82f, 88, 93, 96, 99, 103, 106, 108, 110, 127ff, 146
lexicographic order 39
Lindelöf degree vii, 3, 100ff
linear ordering 53f, 60, 71, 120, 146
measurable 14
network 90ff
nowhere dense 112
nwdc 112ff
one-point gluing 4, 13f
one-point compactification 57, 65, 76, 83
ordinary sup-function 19, 36, 55, 73, 87, 92
patchwork property 9, 11
π-character vii, 2, **76ff**, 119
π-homogeneous 49
π-weight vii, **47ff**
regular space 93
regular ultrafilter 15, 35, 145
right-separated 99ff, 111, 147
rigid 106, 118ff, 125, 131ff, 137
Rubin's algebra 57, 106, 111ff, 123, 134, 137ff, 142
S-space 57
saturated 14, 34, 38, 73, 141
set product 4, 7
sheaf 4, 8ff
Sikorski's theorem 7, 21, 25, 44, 50, 52, 56, 70, 136
singular cardinal 7, 33, 39, 83, 100, 104, 110, 120, 148
Souslin line 22, 46, 55, 93, 102, 104, 130, 146
Souslin tree 16
spectrum 5, 21, 44, 66, 74
spread vii, 2, 4f, 20, 28, **89ff**, 101, 140
strongly inaccessible 22
subalgebra spectrum 5
subdirect product 4, 7ff
superatomic 6, 69ff, 74f, 79, 85, 117, 121
tightness vii, 2, 36, **82ff**, 142, 146
topological density vii, 2, **41ff**
tree 16, 61, 105, 147
ultraproduct 4, 14f, 34, 42, 47, 55, 66, 72, 77, 85, 90, 141
unions of BA's 4, 19, 73, 87, 141
V=L 22, 30, 89
weak products 7, 34, 42, 54f, 76, 82, 94, 117, 119
weakly compact 22, 29, 40, 89
weakly dense 54
weakly inaccessible 7, 22, 105
well-founded 111ff, 123